Aerodynamics for the Professional Pilot

Aerodynamics for the Professional Pilot

Richard Bowyer

Airlife
England

Copyright © 1992 by Richard T. Bowyer

First published in the UK in 1992
by Airlife Publishing Ltd

British Library Cataloguing in Publication Data
A catalogue record for this book is
available from the British Library.

ISBN 1 85310 326 8

All rights reserved. No part of this book may be reproduced
or transmitted in any form or by any means, electronic or
mechanical including photocopying, recording or by any storage and information
retrieval system, without permission from the Publisher in writing.

Printed by Livesey Ltd., Shrewsbury.

Airlife Publishing Ltd
101 Longden Road, Shrewsbury SY3 9EB

CONTENTS

Chapter		Page
	Prologue – Tables of Units and Formulae – Abbreviations – Conversion Charts – Terminology and Definitions	ix
1.	**Basic Mechanics** – Force – Motion – Power	1
2.	**Properties of Gases and the Gas Law** – Pressure and its Units – General Gas Law – Gas Characteristics Equation – Density of a Gas – Effect of Pressure on Density – Effect of Temperature on Density – Relative Density	7
3.	**The Atmosphere** – Composition – ISA (International Standard Atmosphere) – Airspeeds and Measurement of Airspeed (ASI) – Speed of Sound and Mach Number	13
4.	**Aerodynamic Forces** – Drag – Lift and High Lift Devices – Aerodynamic Language – Aerofoils and their development – Stall and its characteristics	27
5.	**Control and Stability** – Aircraft Axes – Control – Stability – Centres of Gravity and Pressure – Design Effects	55
6.	**The Aeroplane in Flight** – Straight and Level Flight – The Climb – The Dive – The Glide – Take-Off and Landing Approach Case – Turning Flight – High Speed Flight – Autorotation – The Spin	69

Foreword

An understanding of the laws of aerodynamics is fundamental to an appreciation of how an aircraft flies and why. This book has been written in a straightforward and concise manner to enable everyone, from the private pilot to the trainee commercial pilot, to understand the theory and practice of aerodynamics. It is not a textbook for engineers, but an attempt to provide accurate easy-to-understand information to all those who need to understand aerodynamics as part of their everyday flying.

Richard Bowyer

Prologue

Tables of Units and Formulae – Abbreviations – Conversion Charts – Terminology and Definitions

TABLES OF UNITS AND FORMULAE

Specific Fuel Consumption .... lbs/hp/hr
Centigrade Heat Unit .... (CHU)
British Thermal Unit .... (BTU)
Pressure .... (P) lbf/ft² on lbf/in²
Volume .... (V) ft³ or m³
Temperature .... (T) °K or °C or °F
Mass .... (wt) lbs

Density .... $\dfrac{\text{wt (mass)}}{\text{vol}} = \dfrac{\text{lb}}{\text{ft}^3} = \dfrac{\text{lbs/ft}^3}{\text{slugs/ft}^3}$

$$\text{wt} = \dfrac{\text{lb}}{32.2} = \text{slugs}$$

International Standard Atmosphere = (ISA)

mB = millibar = 100 newtons/metre²

One Standard Atmosphere:
 14.7 lbs/in²
 1013.25 mB

 29.91 hg
            = Mercury
 760 mm

1 knot = 1.69 ft per sec

13 ft³ = 1 lb of gas

62.5 lb of water = 1 cu ft

6.25 gals of water = 1 cu ft

10 lbs of water = 1 Imperial Gallon

1400 ft lbf = 1 CHU

General Gas Law: $\dfrac{P_1 V_1}{T_1} = \dfrac{P_2 V_2}{T_2}$ for a constant mass

Dynamic Pressure = ½ pV² = stagnation − P static

ABBREVIATIONS

SFC	Specific Fuel Consumption
SAR	Specific Air Range
ANM	Air Nautical Miles
GFC	Gross Fuel Consumption
TOSS	Take Off Safety Speed
FUSS	Flaps Up Safety Speed
ROC	Rate of Climb
TAS	True Air Speed
IAS	Indicated Air Speed
RAS	Rectified Air Speed
EAS	Equivalent Air Speed

CONVERSION TABLES

BAROMETRIC PRESSURE READINGS

Inches	hPa/Mbs	Inches	hPa/Mbs	Inches	hPa/Mbs	Inches	hPa/Mbs	Inches	hPa/Mbs
28·00	948·2	28·60	968·5	29·20	988·8	29·80	1009·1	30·40	1029·5
01	948·5	61	968·8	21	989·2	81	1009·5	41	1029·8
02	948·9	62	969·2	22	989·5	82	1009·8	42	1030·1
03	949·2	63	969·5	23	989·8	83	1010·2	43	1030·5
04	949·5	64	969·9	24	990·2	84	1010·5	44	1030·8
05	949·9	65	970·2	25	990·5	85	1010·8	45	1031·2
06	950·2	66	970·5	26	990·9	86	1011·2	46	1031·5
07	950·6	67	970·9	27	991·2	87	1011·5	47	1031·8
08	950·9	68	971·2	28	991·5	88	1011·9	48	1032·2
09	951·2	69	971·6	29	991·9	89	1012·2	49	1032·5
28·10	951·6	28·70	971·9	29·30	992·2	29·90	1012·5	30·50	1032·9
11	951·9	71	972·2	31	992·6	91	1012·9	51	1033·2
12	952·3	72	972·6	32	992·9	92	1013·2	52	1033·5
13	952·6	73	972·9	33	993·2	93	1013·5	53	1033·9
14	952·9	74	973·2	34	993·6	94	1013·9	54	1034·2
15	953·3	75	973·6	35	993·9	95	1014·2	55	1034·5
16	953·6	76	973·9	36	994·2	96	1014·6	56	1034·9
17	953·9	77	974·3	37	994·6	97	1014·9	57	1035·2
18	954·3	78	974·6	38	994·9	98	1015·2	58	1035·5
19	954·6	79	974·9	39	995·3	99	1015·6	59	1035·9
28·20	955·0	28·80	975·3	29·40	995·6	30·00	1015·9	30·60	1036·2
21	955·3	81	975·6	41	995·9	01	1016·3	61	1036·6
22	955·6	82	976·0	42	996·3	02	1016·6	62	1036·9
23	956·0	83	976·3	43	996·6	03	1016·9	63	1037·3
24	956·3	84	976·6	44	997·0	04	1017·3	64	1037·6
25	956·7	85	977·0	45	997·3	05	1017·6	65	1037·9
26	957·0	86	977·3	46	997·6	06	1018·0	66	1038·3
27	957·3	87	977·7	47	998·0	07	1018·3	67	1038·6
28	957·7	88	978·0	48	998·3	08	1018·6	68	1038·9
29	958·0	89	978·3	49	998·6	09	1019·0	69	1039·3
28·30	958·3	28·90	978·7	29·50	999·0	30·10	1019·3	30·70	1039·6
31	958·7	91	979·0	51	999·3	11	1019·6	71	1040·0
32	959·0	92	979·3	52	999·7	12	1020·0	72	1040·3
33	959·4	93	979·7	53	1000·0	13	1020·3	73	1040·6
34	959·7	94	980·0	54	1000·4	14	1020·7	74	1041·0
35	960·0	95	980·4	55	1000·7	15	1021·0	75	1041·3
36	960·4	96	980·7	56	1001·0	16	1021·3	76	1041·7
37	960·7	97	981·0	57	1001·4	17	1021·7	77	1042·0
38	961·1	98	981·4	58	1001·7	18	1022·0	78	1042·3
39	961·4	99	981·7	59	1002·0	19	1022·4	79	1042·7
28·40	961·7	29·00	982·1	29·60	1002·4	30·20	1022·7	30·80	1043·0
41	962·1	01	982·4	61	1002·7	21	1023·0	81	1043·3
42	962·4	02	982·7	62	1003·1	22	1023·4	82	1043·7
43	962·8	03	983·1	63	1003·4	23	1023·7	83	1044·0
44	963·1	04	983·4	64	1003·7	24	1024·0	84	1044·4
45	963·4	05	983·7	65	1004·1	25	1024·4	85	1044·7
46	963·8	06	984·1	66	1004·4	26	1024·7	86	1045·0
47	964·1	07	984·4	67	1004·7	27	1025·1	87	1045·4
48	964·4	08	984·8	68	1005·1	28	1025·4	88	1045·7
49	964·8	09	985·1	69	1005·4	29	1025·7	89	1046·1
28·50	965·1	29·10	985·4	29·70	1005·8	30·30	1026·1	30·90	1046·4
51	965·5	11	985·8	71	1006·1	31	1026·4	91	1046·7
52	965·8	12	986·1	72	1006·4	32	1026·7	92	1047·1
53	965·1	13	986·5	73	1006·8	33	1027·1	93	1047·4
54	966·5	14	986·8	74	1007·1	34	1027·4	94	1047·8
55	966·8	15	987·1	75	1007·5	35	1027·8	95	1048·1
56	967·2	16	987·5	76	1007·8	36	1028·1	96	1048·4
57	967·5	17	987·8	77	1008·1	37	1028·4	97	1048·8
58	967·8	18	988·2	78	1008·5	38	1028·8	98	1049·1
59	968·2	19	988·5	79	1008·8	39	1029·1	99	1049·5

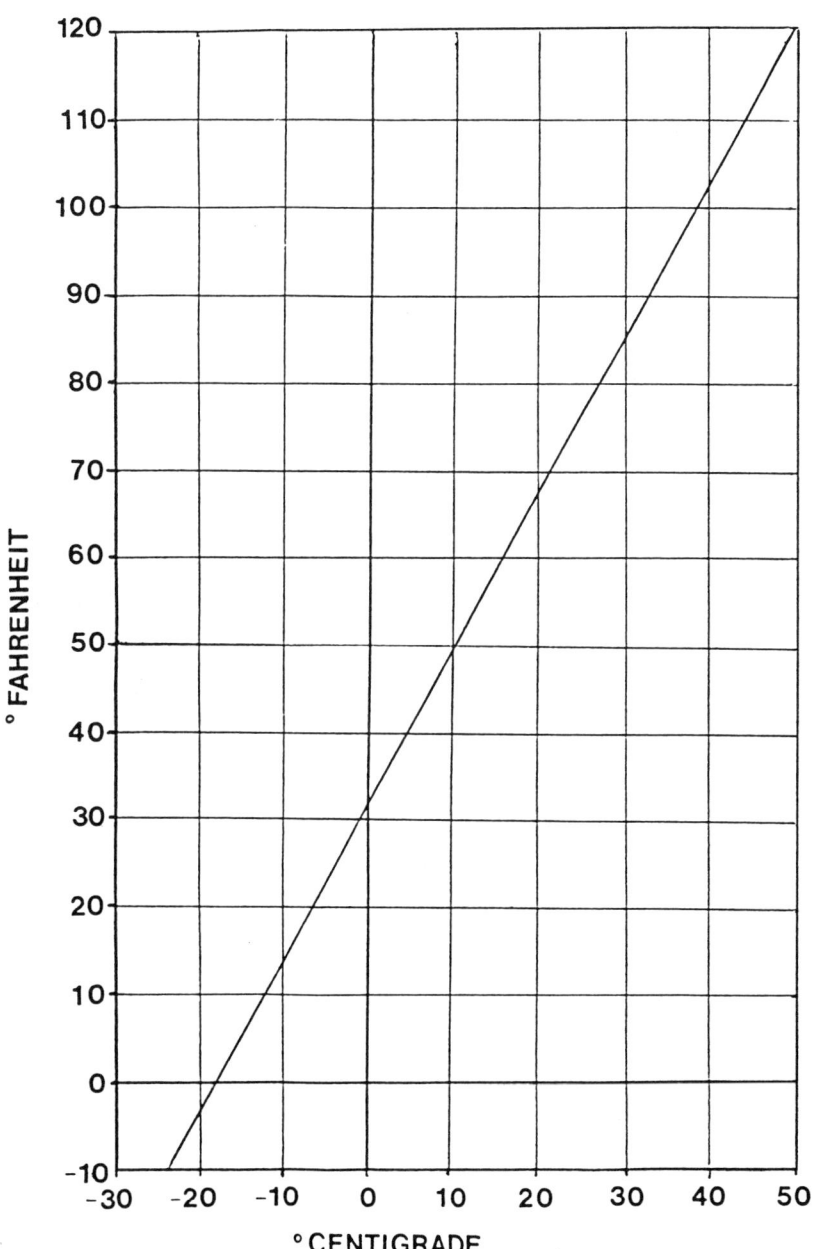

TERMINOLOGY AND DEFINITIONS

Aerofoil — A body so shaped that its motion through the air produces lift without excessive drag.

Angle of Incidence — The angle between the chord line of an aerofoil section and the free airstream direction (commonly known as the ANGLE OF ATTACK).

Aspect Ratio (AR) or (A) — The ratio of the square of the wing span (b^2) to the gross wing area (s).

i.e. $$AR = \frac{b^2}{S}$$

For a wing of rectangular platform

$$AR = \frac{span}{chord} = \frac{b}{c}$$

Centre of Pressure (cp) — The point of intersection of the total air reaction with the chord line. The position of the cp changes with the angle of incidence.

Chord Length — (Chord) (c) – The length of that part of the chord line intercepted by the aerofoil section boundary.

Chord Line — The straight line through the centres of curvature at the leading and trailing edges of the aerofoil section.

Drag (D) — That component of total air reaction, PARALLEL TO the FREE STREAM air flow. (Acts through the cp of course.)

Gross Wing Area — The surface area of the wing bounded by (s) its tips, and leading and trailing edges including the fuselage portion.

Lift (L) — That component of total air reaction at *Right Angles* to the *Free Stream* airflow. (Acts through cp of course.)

Load Factor — The ratio of total lift to aircraft weight.

i.e. $$\text{load factor} = \frac{L}{W}$$

Span (b) — The overall distance between the tips of a wing.

Surface Loading	The *Total Lift* on a wing (L), divided by the gross wing area (s).
Thickness/Chord Ratio (t/c)	The ratio of the maximum thickness (t) of an aerofoil to its chord length (c).
Total Air Reaction	The resultant force due to the pressure distribution around the aerofoil section. (Represented vectorially by a straight line of the appropriate length.)
Wing Load (w)	The total weight of an aircraft (w), divided by the gross wing area (S). i.e. $w = \dfrac{W}{S}$ lbf/ft²

SPEEDS

ASIR	Airspeed Indicator Reading; the uncorrected reading on an airspeed indicator.
IAS	Indicated Airspeed; ASIR corrected only for instrument error.
EAS	Equivalent Airspeed; IAS corrected for position error and compressibility error.
TAS	True Airspeed; the true airspeed of the aeroplane relative to undisturbed air: $EAS/\sqrt{\sigma} = TAS$.
Mach Number	The ratio of TAS to the speed of sound for the ambient conditions.
MMR	Mach Meter Reading; the uncorrected reading on a Mach meter.
V_S	Stall Speed; the speed at which the aeroplane exhibits those qualities accepted as defining the stall.
V_{MS}	The minimum speed in the stall; the minimum speed achieved in the stall manoeuvre.
V_{S1}	The (not more than) zero thrust stall speed at a specified flap setting.
V_{S0}	The (not more than) zero thrust stall speed at the most extended landing flap setting.

V_{NO}/M_{NO}	Normal operating speed; the maximum permitted speed for normal operations.
V_{NE}/M_{NE}	Never exceed speed; a higher maximum permitted speed when operationally desirable.
V_{MO}/M_{MO}	Maximum operating speed; the maximum permitted speed for all operations.
V_{IMD}	Speed for minimum drag.
V_{IMP}	Speed for minimum power.

1.
Basic Mechanics

Force – Motion – Power

1 BASIC MECHANICS

Introduction
The science of aeronautics is based upon the principles of mechanics and energy conversion which are involved in the various forces and reactions which enable an aircraft to fly. It's necessary for aircrews and pilots especially to understand these principles if they are to operate their aircraft effectively. The aim of this chapter is to explain enough about mechanics to achieve this understanding.

Basic Units – The Foot–Pound–Second System
The aeronautical world at present uses the food–pound–second (fps) system of units. In the fps system the basic mechanical units are:
   a. For displacement – the foot (ft)
   b. For mass – the pound (lb)
   c. For time – the second (s).

All other units are derived from the basic units. The required conceptions of displacement and time are simple and no enlargement is needed here, but the idea of mass needs explanation.

Mass
The Law of Inertia – Newton's First Law

A body tends to continue what it is already doing.

> 'A body at rest remains at rest; a moving body tends to continue moving at the same speed and in the same direction'.

In both cases the body is in equilibrium and will stay in that condition unless disturbed by some external force. This tendency to stay in equilibrium is called the inertia of the body. Inertia is a property and cannot be evaluated, we can only measure it in terms of the mass of the body, or the quality of matter in the body.

Momentum
The basic idea of mass does not answer all the problems associated with equilibrium. When motion is involved a new term has to be derived (obtained) from the basic units. The difficulty of stopping a body depends not only on mass but also on velocity. In fact, it depends upon the momentum of the body, that is the product of mass and velocity.

$$\text{Momentum} = \text{Mass} \times \text{Velocity}$$

Force
In order to change the state of equilibrium of a body – to change its momentum – it is necessary to apply a force. Force, then, may be defined as:

> That which changes or tends to change the state of rest or of uniform motion of a body.

Two relationships follow from this definition:

a. When a force is applied the momentum must change, and that:

$$\text{Force} = \text{Rate of change of momentum}$$

b. To change momentum, assuming constant mass, an acceleration must be imposed on the body, and that:

$$\text{Force} = \text{Mass} \times \text{Acceleration}$$

Units of Force
Since Force = Mass x Acceleration, the unit of force must be the units of mass and acceleration. In the fps system this is pound foot per second – which is shortened to poundal. This force of one poundal will impose an acceleration of one ft/sec^2 per second on a mass of one pound. The poundal is small for engineering. The standard unit of force is taken as the force exerted by gravity (under specific conditions) on a mass of one pound. This unit is called the pound force (lbf). When a pound force acts on a pound mass such as a free falling body, the acceleration is 32.2 ft/s^2. It follows that one pound force is equal to 32.2 poundals.

The unit of mass must be 32.2 lb, this unit of mass is called the slug.

Weight
The weight of an object is the force imposed on the body's mass by gravity. Gravitational force, which is the attraction of masses, changes with distance between the masses, of which one is usually the earth, and the other factors, such as the relationship of other planets. Although the mass of a body is constant, its weight may vary considerably, even become zero under some conditions. The expression for force is also true for the force due to its weight.

$$\text{Force} = \text{Mass} \times \text{Acceleration}$$

and

$$\text{Weight} = \text{Mass} \times \text{Gravitational Acceleration}$$

All your work and flying in conventional aircraft is near the earth.

For all practical work, a 20 pound mass is attracted to the earth with a 20 pound force and its weight is 20 pounds. A mass of one slug is attracted to the earth with a force of 32.2 pounds and has a weight of 32.2 pounds.

Turning Effect of a Force – Moment
When an object is at rest it must be supported by a force equal to its weight. When the supporting force and the weight do not act in the same straight line there is a turning effect known as the moment of the force.

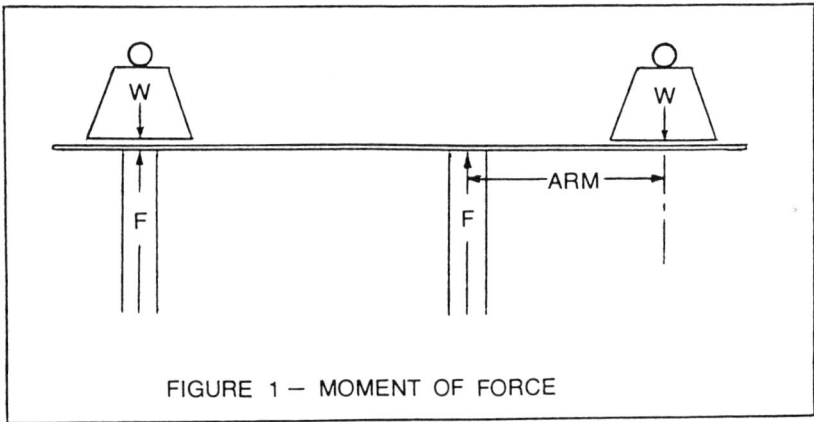

FIGURE 1 – MOMENT OF FORCE

The moment of the force is the product of the force and the perpendicular distance between the force and the weight. This distance is known as the arm.

$$\text{Moment (lbf-ft)} = \text{Force (lbf)} \times \text{Arm (ft)}$$

The moment of a force is the turning effect of a static force. When the force is moving and has a turning effect it is called torque.

Torque
A moving force, such as a hand cranking a car or a piston driving a crankshaft, which has a turning effect produces a torque. The perpendicular distance between the force and the centre of rotation of the revolving shaft is called the effective radius.

$$\text{Torque (lbf-ft)} = \text{Force (lbf)} \times \text{Radius (ft)}$$

Power
Early engines were built to replace horses and work at a rate of the number of horses they replaced. To ensure an engine could

replace and outwork the number of horses it replaced, James Watt measured the amount of work a horse could do in one minute, added a bit on and created a standard for horse-power (hp).

One horse-power is a rate of work:
33,000 ft lbf/min
which is also 550 ft lbf/sec

Example
An aircraft when flying at true air speed of 450 knots requires a thrust of 12,000 lbf. Calculate the thrust horse-power required.

$$1 \text{ knot} = 1.69 \text{ ft/s}$$
$$\text{therefore } 450 \text{ knots} = 450 \times 1.69 \text{ ft/s}$$
$$\text{work done per second} = \text{thrust} \times \text{velocity in ft/s}$$
$$= 12{,}000 \times 450 \times 1.69 \text{ ft lbf}$$

$$\text{Thrust Horse-Power (THP)} = \frac{\text{Work per second}}{550} \text{ hp}$$

$$= \frac{12000 \times 450 \times 1.69}{550} \text{ hp}$$

$$\text{THP} = 16{,}600 \text{ hp}$$

Potential Energy

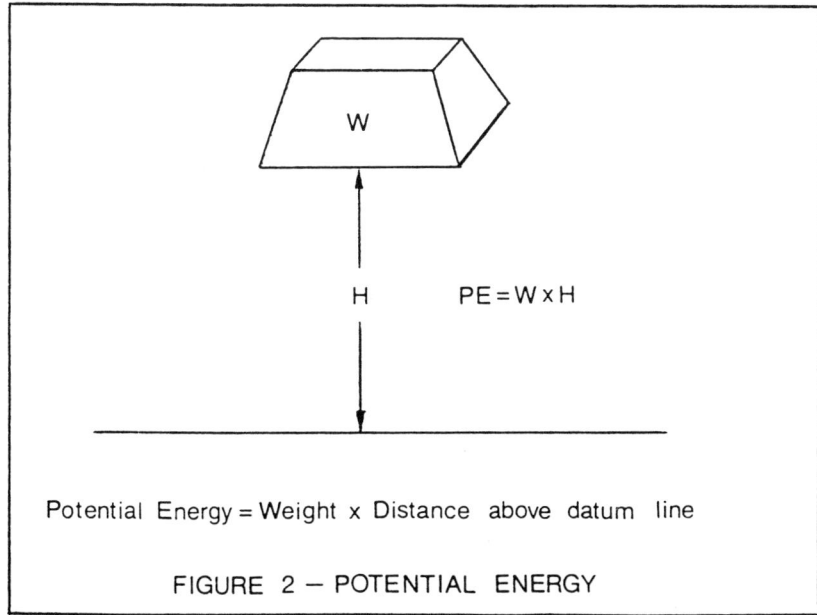

Potential Energy = Weight x Distance above datum line

FIGURE 2 – POTENTIAL ENERGY

This is energy a body possesses due to its position in relation to a suitable level. In a factory the level may be the floor. An aircraft in flight the level may be mean sea-level, but whatever level is chosen it is suitable for the work in hand. When a body is lifted from the reference level, known as a datum line, to its position above it, work is done in lifting it. This work is not lost but stored in the body as *Potential Energy* (PE).

Kinetic Energy
This is energy due to the velocity of the body, and when it falls from rest loses potential energy. Energy cannot be lost so this energy must appear as another form. As the body falls it gains velocity and the energy becomes *Kinetic Energy* (KE).

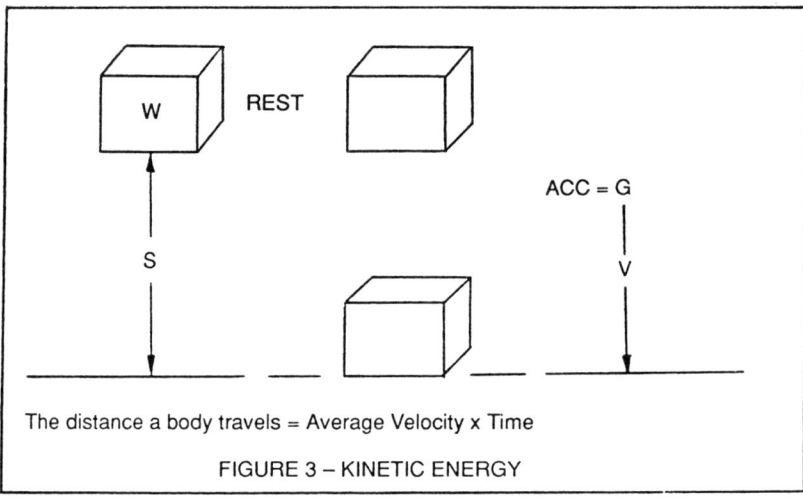

The distance a body travels = Average Velocity x Time

FIGURE 3 – KINETIC ENERGY

The Law of Acceleration — Newton's Second Law
In his second law, Newton compensates for the unbalanced force mentioned in his first law. If there is an unbalanced force on the body, it accelerates. Newton stated: 'The acceleration is directly proportional to the force, inversely proportional to the mass and in the direction of the unbalanced force'.

$$a = \frac{F}{M}$$

The Law of Reaction – Newton's Third Law
'For every action there is an equal and opposite reaction'.

The jet engine operates on the principle of the third law. A forward force on the engine, called thrust, is produced as air is accelerated rearwards through the engine.

2.
Properties of Gases and the Gas Law

Pressure and its Units – General Gas Law – Gas Characteristics Equation – Density of a Gas – Effect of Pressure on Density – Effect of Temperature on Density – Relative Density

2 PROPERTIES OF GASES AND THE LAW

Introduction
We are concerned with the following properties:
  a. Pressure (P)
  b. Volume (V)
  c. Temperature (T)
  d. Mass (W)
  d. Density (ρ) (pronounced RHO).

The pressure is the force exerted by the gas on unit area. Ordinary pressure gauges measure pressure in lbf/in^2 above atmospheric pressure. When there is any doubt the units are written – lbf/in^2 g, indicating it is a gauge pressure.

Absolute Pressure
When calculating the behaviour of gases it is necessary to know 'all' or the 'absolute' pressure exerted by the gas. Absolute pressure is gauge pressure plus standard atmospheric pressure.

Standard (SA) sea-level atmospheric pressure is 14.7 lbf/in^2: to convert a gauge pressure to absolute pressure:
Gauge pressure (lbf/in^2 g) + 14.7 = absolute pressure (lbf/in^2 abs).

Units of Absolute Pressure
The units of absolute pressure often depend upon the way the pressure is measured. A standard atmospheric pressure of 14.7 lbf/in^2 abs will support a column of Mercury (Hg):

	760 mm high
this is equal to	29.91 in Hg
and equal to	1013.25 milli Bars (mB)

A pressure of one millibar is a force of 100 Newtons per square metre. This is an unusual unit but it is essential that you remember:

ISA sea-level atmospheric pressure is 1013.25 mB

Volume
The volume of a contained gas is the volume of its container. The

volume of a moving gas is the volume occupied by a given weight. The volume of a pound of gas is known as specific volume. The unit of volume is the cubic foot.

Temperature
The temperature scale used for gas calculations is the Kelvin scale (°K).

Mass
Mass is measured by weight; the symbol 'W' is used for mass and weight in pounds. It is convenient in some density calculations to have mass in slugs.

$$\text{Mass in slugs is } \frac{W}{g}$$

Density
The density of a substance is its mass per unit volume.

$$\text{Density} = \frac{\text{mass}}{\text{volume}} = \frac{W}{V} \text{ lb/ft}^3$$

Now as $\frac{W}{g}$ is mass in slugs,

$$\text{Density} = \frac{\text{mass}}{\text{volume}} = \frac{W}{gV} \text{ slugs/ft}^3$$

Properties and Units
a. pressure   P in lbf/ft² abs.
b. Volume   V in ft³
c. Temperature   T in °K
d. Mass   W in lb and $\frac{W}{g}$ in slugs
e. Density ρ in lb/ft³ when mass is W and ρ in slugs/ft³ when mass is $\frac{W}{g}$

General Gas Law
This law shows the relationship of pressure, temperature and volume of a given mass of gas. Robert Boyle in 1662 showed, for a given mass of gas at constant temperature, that:

Pressure is inversely proportional to the volume.

$$P \times V = \text{constant}$$
$$\text{and } P_1 V_1 = P_2 V_2$$

J. A. C. Charles (1746–1823) showed that all gases have the same coefficient of expansion. This work, expressed by William Thomson (1824–1907) later Lord Kelvin, gives the relationship, for a given of gas constant pressure, that:

Volume is proportional to the Absolute Temperature

$$\frac{V}{T} = \text{Constant}$$

and

$$\frac{V_1}{T_1} = \frac{V_2}{T_2}$$

It is impossible to keep the temperature or pressure constant during a practical change all properties will change simultaneously. It is necessary to have a relationship between pressure, temperature and volume. Therefore:

$$\frac{P_1 V_1}{T_1} = \frac{P_2 V_2}{T_2}$$

This is the *General Gas Law.*

Gas Characteristics Equation

To find the density of a substance we must know its mass. It is not easy to weigh a gas, but if the weight can be expressed in terms of other properties, which can easily be measured, then the density can be found. The General Gas Law states:

$$\frac{PV}{T} = \text{constant for a given mass}$$

By taking a mass of 1lb the constant can be found for a gas. The volume will be the volume of one pound. The units of the constant will be the units of the equation.

$$\frac{P \text{ lbf}}{\text{ft}^2} \times \frac{V \text{ ft}^3}{\text{lb}} \times \frac{1°K}{T} = \text{constant (R)}$$

units of R are ftlbf/lb – °K.

R is known as the gas characteristic constant.
Gas characteristic constant for air – R = 96 ftlbf/lb°K.

Now

$$\frac{P \times (\text{volume of 1 lb})}{T} = R$$

Multiply each side by weight (W)

Now
$$\frac{P \times (\text{volume of 1 lb}) W}{T} = WR$$

but (volume of 1 lb) × W = total volume (V)

hence
$$\frac{PV}{T} = WR$$

and
$$PV = WRT$$

This is the 'Gas Characteristic Equation'.

Density of a Gas

The main use of the gas characteristic equation at present is to find the density of the air.

Now
$$\text{Density} = \frac{\text{weight}}{\text{Volume}}$$

from
$$PV = WRT$$
$$\text{density } (\rho) = \frac{W}{V} = \frac{P}{RT}$$

Effect of Pressure on Density

Consider temperature to be constant then:

$$\rho_1 = \frac{P_1}{RT}$$

and $\rho_2 = \dfrac{P_2}{RT}$ therefore $\dfrac{\rho_1}{\rho_2} = \dfrac{P_1}{P_2}$

and $\rho_2 = \dfrac{P_2}{P_1} \rho_1$

The pressures are a ratio and must be absolute pressures with the same units.

Density Increases with Pressure

Effect of Temperature on Density

Consider pressure to be constant then:

$$\rho_1 = \frac{P}{RT_1} \quad \text{and} \quad \rho_2 = \frac{P}{RT_2}$$

therefore $\dfrac{\rho^1}{\rho_2} = \dfrac{T_2}{T_1}$ and $\rho^2 = \dfrac{(T_1)}{(T_2)} \rho^1$

Density Decreases with Increase in Temperature

Relative Density

Relative density is the relationship of actual density to a chosen standard density. The standard density most generally used is the density of an atmosphere at 14.7 lbf/in^2 and +15°C. This is ISA at sea level and is expressed as ρo.

$$\text{Density } (\rho) = \frac{P}{RT}$$

$$\text{Actual density } (\rho a) = \frac{Pa}{RTa}$$

$$\text{Standard density } (\rho o) = \frac{Po}{RTo}$$

$$\text{Relative density} = \frac{\text{actual density}}{\text{standard density}}$$

$$\rho \text{ rel} = \frac{\rho a}{\rho o}$$

$$\rho \text{ rel} = \frac{Pa}{RTa} \times \frac{RTo}{Pa}$$

Therefore
$$\rho \text{ rel} = \frac{(Pa)}{(Po)} \frac{(To)}{(Ta)}$$

3.
The Atmosphere

Composition – ISA (International Standard Atmosphere) – Airspeeds and Measurement of Airspeed (ASI) – Speed of Sound and Mach Number

3 THE ATMOSPHERE

Introduction
The operation of all aircraft using fixed or moving wings to generate lift depends to a large extent on the properties of the air in which they fly. It is therefore necessary to appreciate the properties and peculiarities of the atmosphere in order to study the performance of aircraft. The aim of this section is to discuss those aspects of the atmosphere which are important to aircrew in the operation of their aeroplane.

Composition
The atmosphere is a thin, fluid skin surrounding the earth and extending out to about 500 miles. It is a mixture of many gases, but for practical purposes may be considered to be four-fifths of nitrogen and one-fifth of oxygen by volume. The pressure and density of the air decrease with height; the water vapour content depends on the temperature which itself decreases with height up to about 36,000 feet, then remains substantially constant up to about 40 miles.

About half the total weight of air is contained in the first 18,000 feet, and a further quarter in the next 18,000 feet. A height of about 18,000 feet represents the danger limit for a human pilot – he *must* be fed on oxygen above this height, but it is of interest to note that in many modern high performance aircraft, the rate of climb is so high and the time to 'oxygen height' so short that aircrew take off and remain on full oxygen. Above about 100,000 feet there is insufficient oxygen to support combustion in a modern jet engine, and flight above this level necessitates the use of rocket engines.

The region close to the earth in which the temperature decreases with height has been given the name *Troposphere* (literally means 'turn sphere'); the region above the troposphere, where temperature is constant, is called the *Stratosphere.* The transition region between them is called the *Tropopause.* The level of the tropopause varies from about 30,000 feet at the poles, to about 54,000 at the equator; in temperate latitudes it lies at about 36,000 feet. This variation in tropopause level with latitude means that the lowest temperatures are encountered at altitudes over tropical areas – which give rise to the apparently contradictory situation of low temperature trials being carried out in the tropics!

Although it is normally considered that there are four different layers of atmosphere there are in actual fact five which are as follows:

Troposphere
Stratosphere
Mesosphere
Ionosphere
Exosphere

We will only be mainly considering the first two as these are the layers in which conventional aircraft normally fly and operate, although the Ionosphere will be discussed with reference to radio waves in the appropriate subjects in another volume.

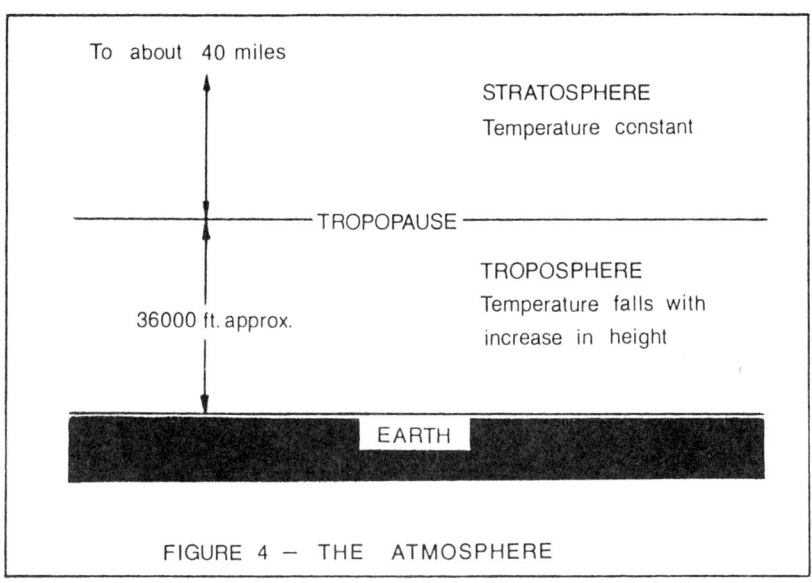

FIGURE 4 — THE ATMOSPHERE

International Standard Atmosphere (ISA)

To enable comparisons to be made between the performance of different aircraft in various atmospheric conditions, a hypothetical atmosphere has been compiled based on world averages. This atmosphere is called the 'International Standard Atmosphere' and performance data are converted to ISA values before comparison with data from other aircraft.

A table of ISA values for various properties of the atmosphere is given along with graphs showing the variation of temperature, pressure and density in the atmosphere.

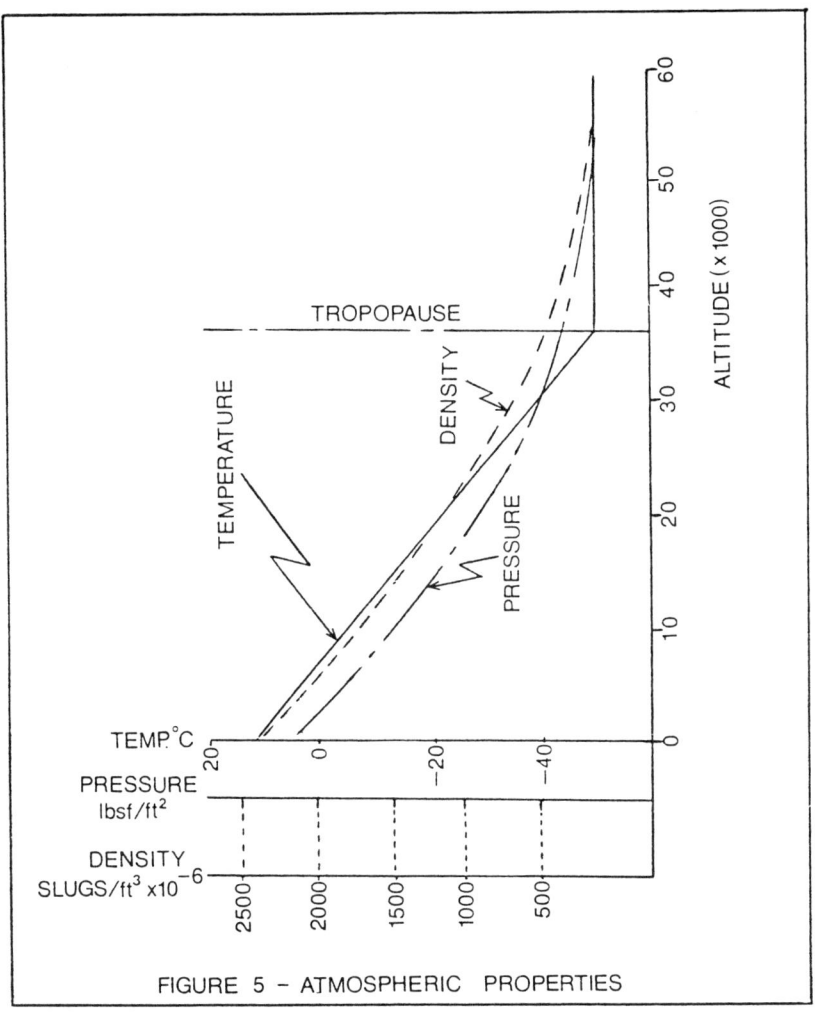

FIGURE 5 – ATMOSPHERIC PROPERTIES

Airspeeds

Aircraft generate lift by moving lifting surfaces through the air. It follows that a knowledge of the relative motions of a wing and air is valuable, and we shall investigate the various airspeed terms and determine their relationships to each other.

The speed at which the free stream air moves relative to the aeroplane is called the *True Airspeed* of the aircraft (TAS). To navigate an aeroplane, the pilot points it in a chosen direction (called the 'heading') and flies it in such a way as to achieve a certain true airspeed. The combination of the TAS/hdg vector with the wind vector determines the *Track* and *Groundspeed* of the aircraft as illustrated in Figure 6.

International Standard Atmosphere

Altitude 1000's of ft	Temperature °C T	Pressure lbf/ft² P	Density slugs/ft³ multiply by 10⁻⁶	Relative Density ρ rel	$\frac{1}{\sqrt{\rho\ rel}}$	Speed of sound knots Vs
0	15.0	2116	2378	1.000	1.00	661
5	5.1	1760	2049	0.862	1.08	650
10	−4.8	1455	1756	0.738	1.16	638
15	−14.7	1194	1496	0.629	1.26	626
20	−24.6	972	1267	0.533	1.37	614
25	−34.5	785	1065	0.448	1.49	602
30	−44.4	628	889	0.374	1.63	589
35	−54.3	498	736	0.310	1.79	576
36,090	−56.5	452	671	0.284	1.88	573
40	−56.5	392	582	0.246	2.02	573
45	−56.5	307	458	0.193	2.37	573
50	−56.5	242	361	0.152	2.56	573
55	−56.5	196	284	0.119	2.89	573
60	−56.5	150	224	0.094	3.26	573
65	−56.5	118	176	0.074	3.68	573

Note: One knot = 1.69 ft/sec
Temperature lapse rate = 1.98° C per 1000 ft up to tropopause

The standard sea-level conditions for the ISA are seen to be:
a. Temperature − 15° C or 288° K
b. Pressure − 2116 lbf/ft² (14.7 lbf/in², 1013.25 millibars)
c. Density − 0.002378 slugs/ft³
c. Speed of sound − 661 knots.

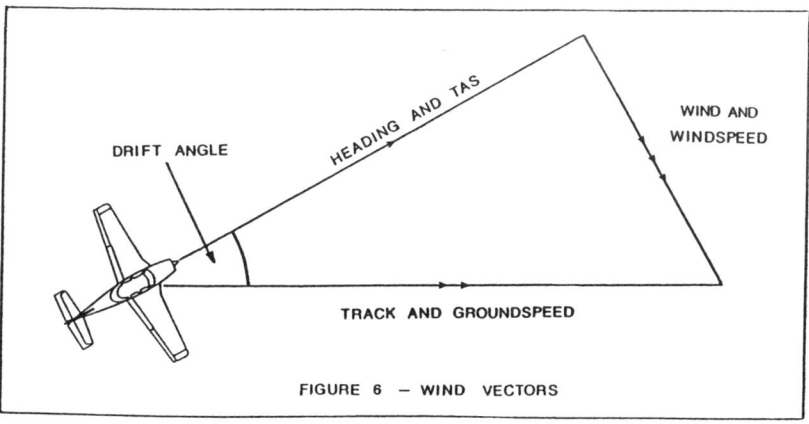

FIGURE 6 − WIND VECTORS

Wind and Windspeed

When the air changes temperature it will also have changes in pressure (in accordance with the gas laws) which result in movement of the air, the direction of movement always being from high pressure to low pressure regions. Such air movements are called winds, although the word wind is usually associated with movement of air in a horizontal direction across the earth's surface. In addition, the *convection* currents associated with clouds and thunderstorms may produce strong *vertical* winds often referred to as *gusts* or *thermals* which may be as fast at 60 mph, and can be used to gain height when soaring. Gliding for long distances relies upon such thermals to restore height periodically.

There is often a strong wind *gradient* near the ground, due to *skin-friction* or surface roughness, the speed of the wind increasing with height, and in addition due to the rotation of the earth, the direction of the wind also varies with height, it tending to veer clockwise in the Northern Hemisphere.

Motion of the air itself will obviously affect the overall behaviour of an aircraft, so it is necessary to be quite clear about certain definitions in relation to speed and direction.

Heading and Airspeed

The *Heading* or course to steer, is the direction the aircraft is pointing.

The speed of the aircraft *relative* to the air is known as the *Airspeed*, and is measured by the airspeed indicators (ASI).

Both *Heading* and *Airspeed* are under control of the pilot.

Track and Groundspeed

The resultant of the *Airspeed* along the heading, and the *Windspeed* is the speed of the aircraft relative to the ground, (the *Groundspeed*) and the path of the aircraft over the ground is called the *Track*, (Figure 6).

In *Still* air, *Airspeed* and *Groundspeed* will be equal. At other times for a given airspeed, the groundspeed will be increased by a tailwind, or reduced by a headwind. Thus landing and take-off are performed *into* wind, in order to reduce the groundspeed required, and the distance in contact with the ground. Landing speeds of carrier-borne aircraft are further reduced by the carrier steaming into the wind.

Of all the information presented to the pilot by his instruments, the airspeed of the aircraft, and its height above sea-level are probably the most important. It will be seen later that when the aircraft speed approaches the speed of sound, the relative values of these two parameters are also important.

The Measurement of Airspeed
Airspeed is measured by a pressure instrument, called an airspeed indicator (ASI). It gives an indication of the speed of the aircraft relative to the air, and will read ground speed *only in still air at sea-level.*

The ASI is made up of two assemblies:
a. The pitot-static (or pressure) head.
b. The instrument box.

The Pitot-Static Head
This consists of two separate tubes, one of which has an open end facing directly into the airflow (the pitot) and the other of which has a closed end but has holes positioned around its circumference. The pitot tube is often fitted inside the static tube as in Figure 7 below:

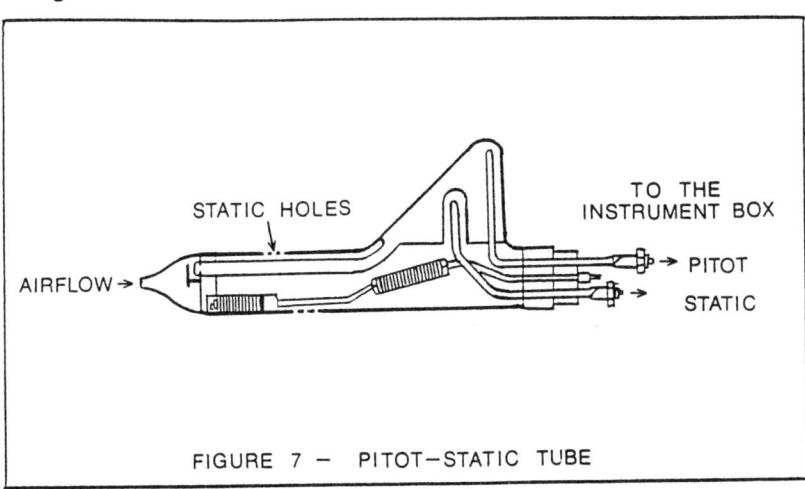

FIGURE 7 — PITOT-STATIC TUBE

The Instrument Box
This consists of a metal capsule fed with pitot air, within a box which is fed with static air. The capsule is arranged so as to drive a pointer over a scale via a suitable linkage. It will be readily appreciated from the following diagram (Figure 8) that the pointer movement will depend on the difference in pressures between the pitot air and the static air.

Errors in the ASI
There are several sources of error in the ASI, each one of which may be more or less significant depending on the condition of flight concerned, e.g. high or low speed, high or low altitude. We shall deal with 'instrument', 'position', 'compressibility' and 'density' errors.

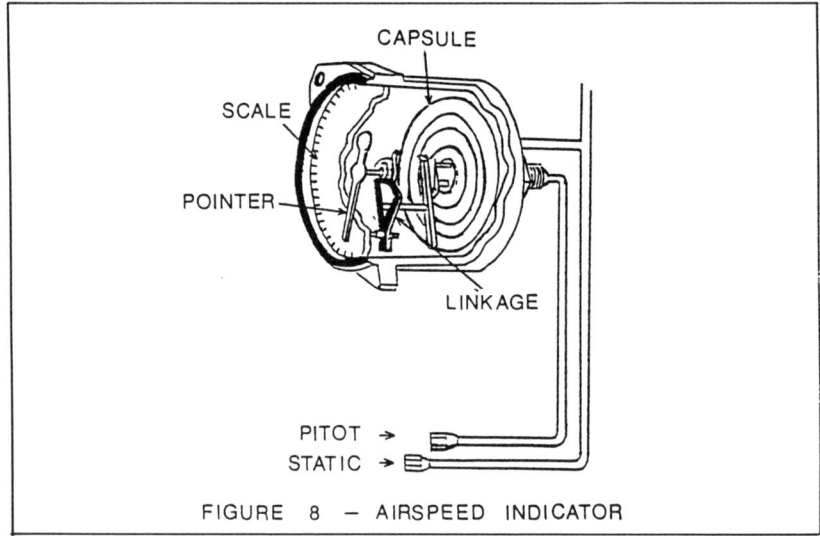

FIGURE 8 — AIRSPEED INDICATOR

Instrument Error
Airspeed indicators are manufactured by mass production techniques and it is inevitable that differences in performance will result. Each instrument is compared with a standard and rejected if its inaccuracy is beyond specified limits. The basic reading given by an ASI is called the airspeed indicator reading (ASIR); it must be corrected for instruments error to give indicated airspeed (IAS). Instrument errors are usually small enough to be negligible.

Position Error
Because of the attitude of the aircraft or the location of the static source, the static pressure supplied to the ASI may not always be the true ambient pressure. Every aircraft type has exhaustive flight tests carried out on it to determine the best position for the static source; sometimes the source will be separate from the pitot tube. The error introduced is known as position error and graphs of PE corrections are produced by manufacturers for various configurations (flaps up or down, gear up or down), height and speed ranges. Typical ASI position error curves are shown in Figure 9. When ASI has been corrected for position error (which is sometimes known as pressure error) it is known as *Rectified Airspeed* (RAS).

Compressibility Errors
Above about 200–300 knots, the effects of compressibility becomes appreciable and cause the ASI to *over-read.* Correction tables are produced and the RAS is adjusted from them to give *Equivalent Airspeed* (EAS).

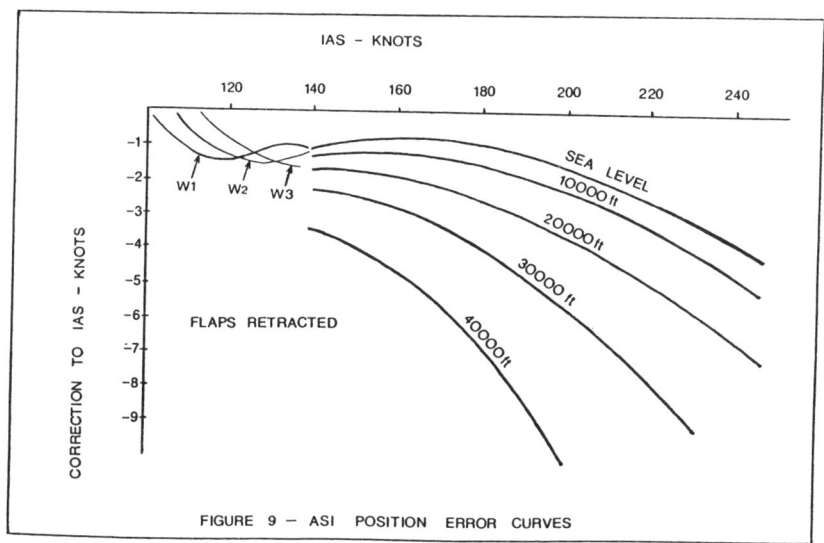

FIGURE 9 — ASI POSITION ERROR CURVES

Environmental Problems

Errors may be caused in the ASI by such outside influences as dirt and ice. For these reasons pitot-static heads are provided with covers and static sources with plugs (both *must* be removed before flight), and pitot heads are heated during flight in icing conditions.

Error Table

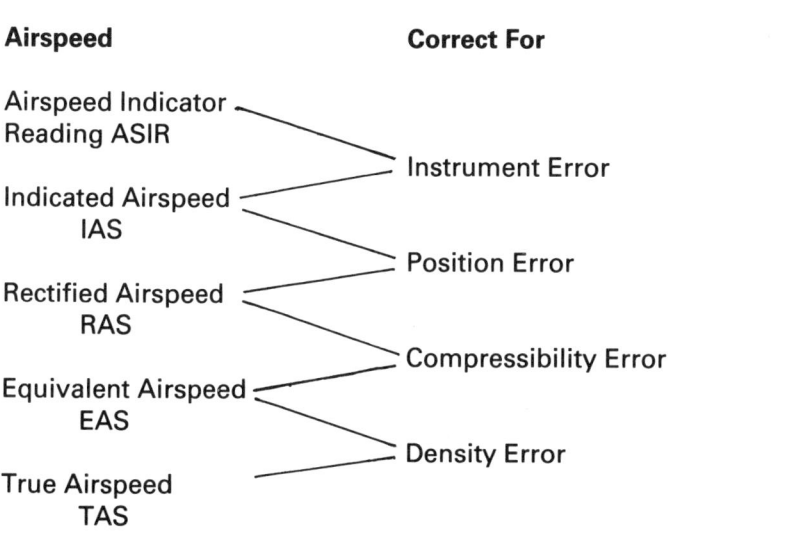

Flow Energy

In the tube shown in Figure 10, the force on a given section is equal to Pressure x Area.

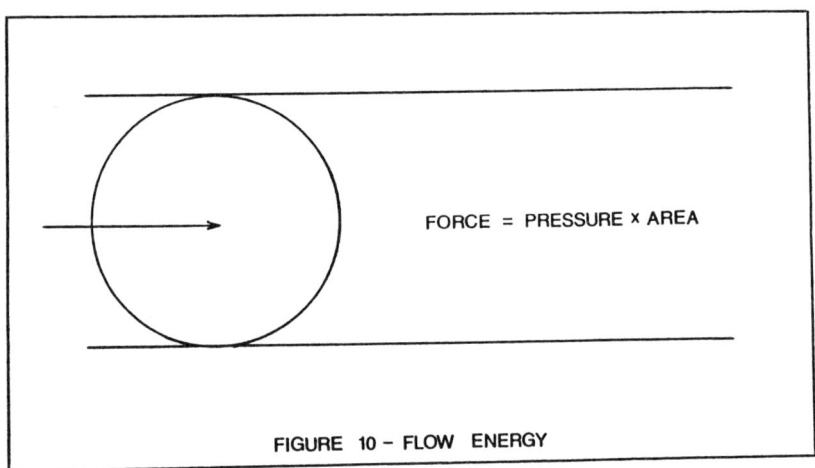

FIGURE 10 - FLOW ENERGY

Work done = Force x Distance, and if unit time is considered,

WD = Force x Velocity
   = Pressure x Area x Velocity
   = Pressure x Volume Flow

and for a unit mass, volume = specific volume

Flow energy = Pressure x Specific Volume
(ft lbf)   (lbf/ft^2)   (ft^3)

Steady Flow Energy Equation

This is more commonly known as *Bernoulli's Equation,* and it states that for a flow of an incompressible, non-viscous fluid that:

'At all points along a given Streamline, the sum of the Pressure Energy, the Kinetic Energy and the Potential Energy is Constant.'

(A streamline is the path of a *given* particle in the fluid flow) so, along any streamline,

Pressure Energy + Kinetic Energy + Potential Energy

= *Constant*

But, in aerodynamics, changes in *Potential* energy along a streamline are sufficiently small to be neglected, so that for all practical purposes, *Bernoulli's Equation* becomes:

Pressure Energy + Kinetic Energy = Constant (Along a streamline)

Mathematically this is the same as saying that:

$$Pa + \tfrac{1}{2} \rho V^2 = \text{constant},$$

where

Pa = static air pressure (lbf/ft²)
ρ = air density (slugs/ft³)
V = air velocity (ft/sec)

Thus at two *different* points 1 and 2 along the *same* streamline we have:

$$Pa_1 + \tfrac{1}{2} \rho V_1^2 = Pa_2 + \tfrac{1}{2} \rho V_2^2$$

If now at the point 2 the moving airstream is brought to rest, that is $V_2 = 0$, then the point 2 is called a *Stagnation Point*, and the expression above becomes:

$$Pa_1 + \tfrac{1}{2} \rho V_1^2 = Pa_2^*$$

where Pa_2^* denotes what is called the *Stagnation Pressure*, which of course is the same at all points on a *given* streamline.

The difference between the *stagnation* pressure and the *static* pressure at any point gives the *Dynamic Pressure* at that point, i.e.

$$Pa_2^* - Pa_1 = \tfrac{1}{2} \rho V_1^2$$

where $\tfrac{1}{2} \rho V_1^2$ is the *Dynamic Pressure* at the point 1.

Now, since the pitot tube is open to the direct airflow, the moving airstream passes into the capsule, and it is brought to rest. Thus *inside* the capsule the pressure is the stagnation pressure Pa_2^*.

The static tube allows air at atmospheric pressure (free-stream static pressure) to pass into the instrument box and surround the capsule.

The *Difference* in pressure across the capsule, i.e. that causing movement of the pointer is the *difference* between the stagnation pressure and the static pressure. Thus pressure difference across the capsule is given by:

$$Pa_2^* - Pa_1 = (Pa_1 + \tfrac{1}{2} \rho V_1^2) - Pa_1$$

$$= \tfrac{1}{2} \rho V_1^2 \text{ (the dynamic pressure)}$$

Changes in airspeed will cause changes in dynamic pressure which will result in the pointer of the instrument moving the dial. Provided the air density is known, by suitable calibration, the dial of the instrument can be made to read the speed of the moving airstream V_1. (This is of course the same as the aircraft airspeed.)

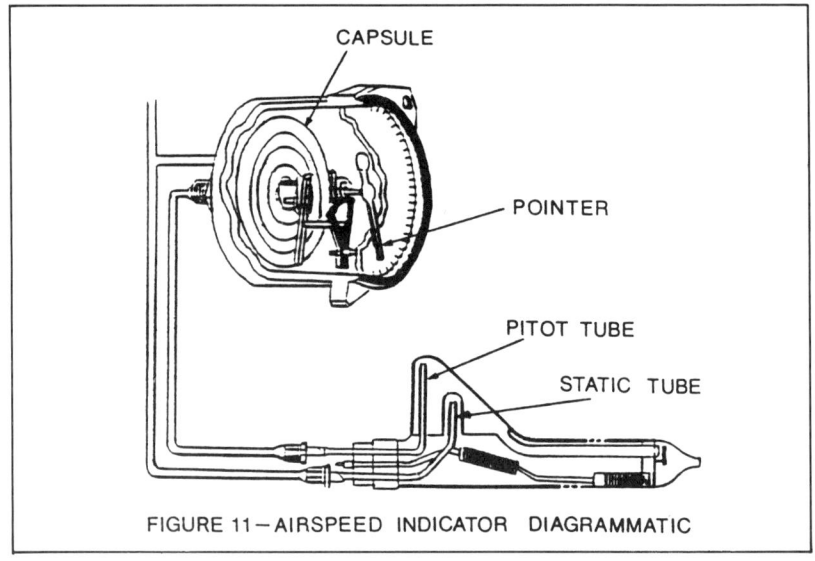

FIGURE 11 — AIRSPEED INDICATOR DIAGRAMMATIC

Thus the diaphragm which operates the pointer has Pa on one side and Pa + ½ ρ V² on the other; the result is ½ ρ V², the dynamic pressure (Figure 11).

The instrument can now be calibrated using air of a known (and constant) density, but the calibration will be true *only in conditions where the air density is that at which the instrument was calibrated.* In fact the density used for calibration is the ISA sea-level value and the ASI is simply a pressure gauge calibrated in airspeed at sea-level density.

For most purposes the ASIR is taken to be the IAS (instrument error is very small) and some types of navigational computer have built-in corrections for compressibility error. It is important to note also that the airspeed which imposes the loads on an aircraft is the EAS (which equals IAS if position and compressibility errors are neglected). This explains why some aircraft are restricted to say, 380 knots TAS at sea-level, but can safely fly at 600 knots TAS at high altitude, the EAS will be the same in each case.

Speed of Sound and Mach Numbers

The local speed of sound in air is extremely significant. At low speeds the particles of air are displaced smoothly by an aircraft because the particles have time to adjust to the transient situation. When the airspeed approaches the speed of sound the particles cannot move easily in time and are more violently displaced. This results in abrupt changes of pressure, temperature and density, which produce large increases in drag and other undesirable effects.

The ratio of true airspeed to the local speed of sound is known as the *Mach Number*, (M), after Ernst Mach, the Austrian scientist famous for his early study of high speed aerodynamics.

Thus $M = \dfrac{TAS}{V_s}$

where V_s is the local speed of sound and the speed of sound is found to vary only with absolute temperature. It follows that the speed of sound falls with increase in altitude until the tropopause is reached, thereafter remaining substantially constant.

The Machmeter
In order that aircrew should have an indication of Mach Number, high speed aircraft are provided with a Machmeter which gives a direct reading of flight Mach Number.

The instrument consists of an airtight box containing two capsules – one similar to the ASI capsule and one similar to the altimeter capsule. Both capsules are linked to the pointer and it can be shown that a suitable linkage will give a direct reading of Mach Number.

Because of its similarity to the ASI and the Altimeter, the Machmeter will be subject to the same errors. Position error is usually small. Density correction is unnecessary because it is absorbed in the linkage. Instrument errors vary between individual instruments and may be large, so careful checking is necessary and Machmeters must be rejected if they are found to be outside the calibration limits.

It will have been seen from all that has previously been discussed that there are different types of fluid flow. We have mainly discussed *Steady Flow* and there are five basic parameters:

a. Direction of Flow
b. Speed of Flow
c. Pressure of Flow
d. Temperature of Flow
e. Density of Flow

Unsteady Flow
It is opposite of steady flow, and in it the flow parameters are varient, and eddies or turbulence is formed.

Rotational Flow
The flow in which is circular, or rotational like in a vortex (i.e. vortices formed at the wing tips).

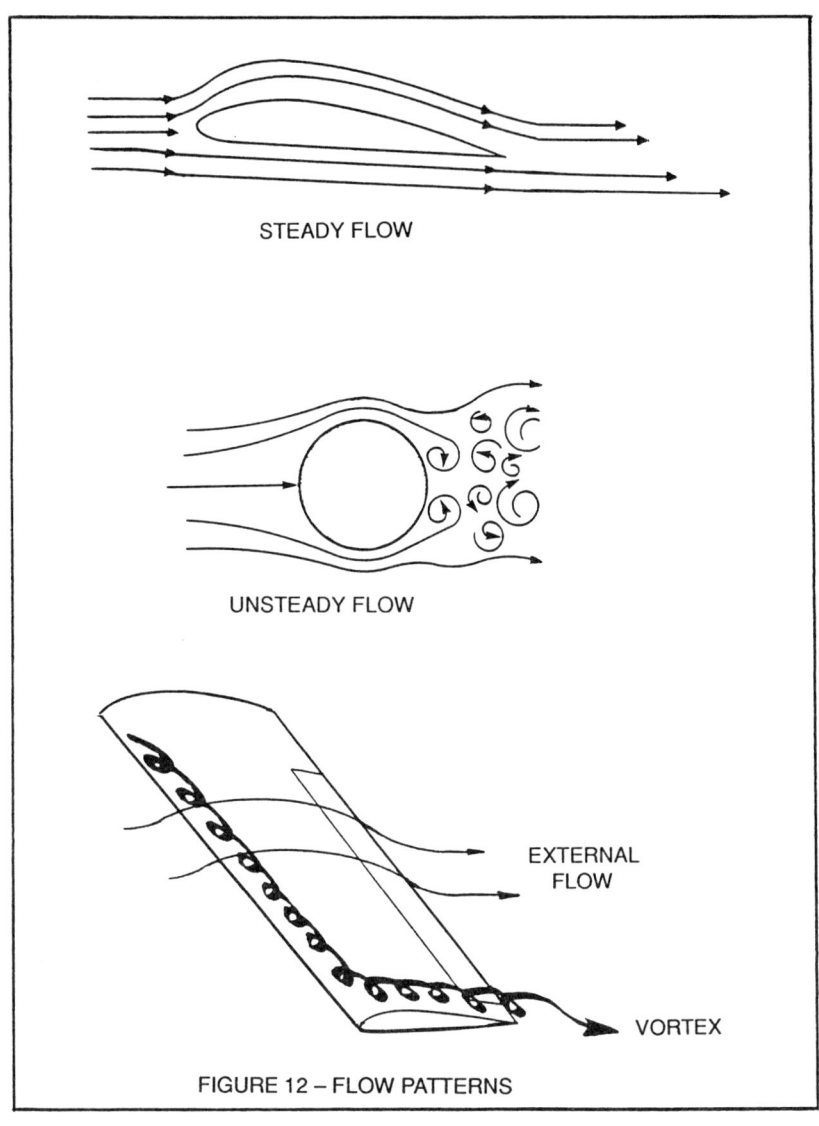

FIGURE 12 – FLOW PATTERNS

4.
Aerodynamic Forces

Drag – Lift and High Lift Devices – Aerodynamic Language – Aerofoils and their development – Stall and its characteristics

4 AERODYNAMIC FORCES

Introduction

When an aeroplane moves through a mass of air, the particles of air are deflected in all directions. The displacements may be translational (with no spin of any sort) or rotational (spin about one or more axes), or a combination of both. Both translation and rotation of a particle require work to be done on that particle, which means that energy must be expended by the aeroplane. (The source of the energy is the aircraft's fuel.) The action of the aircraft on the air particles produces reaction of air particles on the aircraft. This 'total air reaction', as illustrated in Figure 13, is the source of the aerodynamic forces on the aircraft, and is normally separated into two components:

a. Drag – the component of total air reaction which acts parallel to the free-stream airflow
b. Lift – the component of total air reaction which acts at right angles to the free-stream airflow.

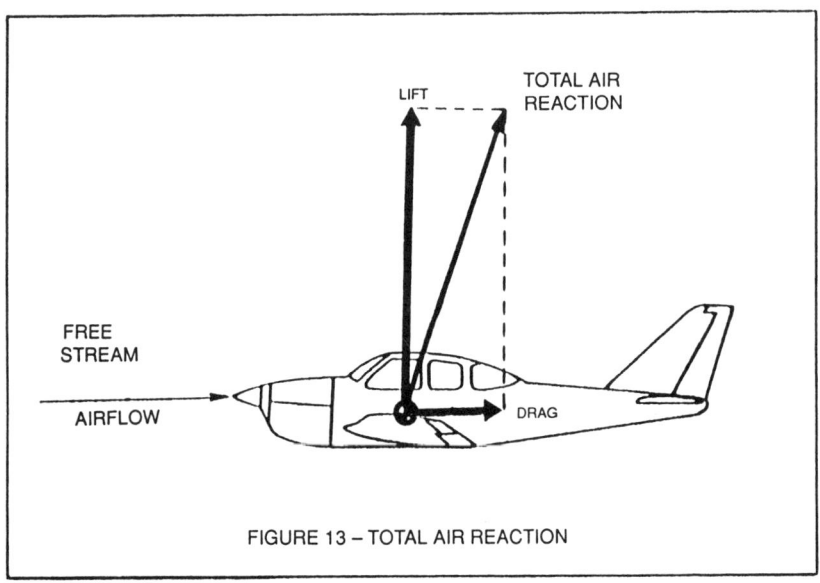

FIGURE 13 – TOTAL AIR REACTION

The aim of these notes is to study drag and lift in detail and to develop relationships between them and the other qualities such as air density, airspeed, etc. We shall also consider the effects of aircraft configuration on the two forces.

Drag

Total Drag
The total drag of a low speed aircraft can be shown to be comprised of two components:

    a. Profile (or zero-lift) drag
    b. Induced (or lift-dependent) drag.

In both cases the drag concerned must be matched by corresponding forward forces, which in straight and level flight will be provided by the thrust of the engine(s).

Profile Drag
The drag of an aircraft which is contributed by its physical shape and condition is called profile drag, and may be sub-divided into two kinds:

    a. Form Drag (the shape)
    b. Skin Friction (the condition, i.e. rough or smooth).

Profile drag is directly proportional to dynamic pressure ($\frac{1}{2} \rho V^2$), and it depends on physical size (or area, S) and on shape and condition (i.e. whether or not a body is streamlined, which brings in a shape factor, or 'drag coefficient' C_{Dp}). The situation can be expressed graphically or in the form of an equation:

$$D_p = C_{Dp} \tfrac{1}{2} \rho V^2 S$$

where D_p is the profile drag in lbf

    C_{Dp} is a dimensionless force coefficient, called the coefficient of profile drag
    ρ is the density in slugs/ft^3
    V   is the true air speed in ft/Sec
    S   is the suitable area in ft^2 (taken as the gross wing area for complete aircraft).

Form Drag
The contribution towards total drag is a function, basically, of the shape and is called form drag. If a large flat plate is taken at 100% standard, i.e. one at which the relative airflow is brought to rest, then a sphere of the same frontal area is found to have about 60% of the form drag of the plate. A streamlined shape of the same frontal area may produce as little as 2–5% of the flat plate figure as illustrated in Figure 14.

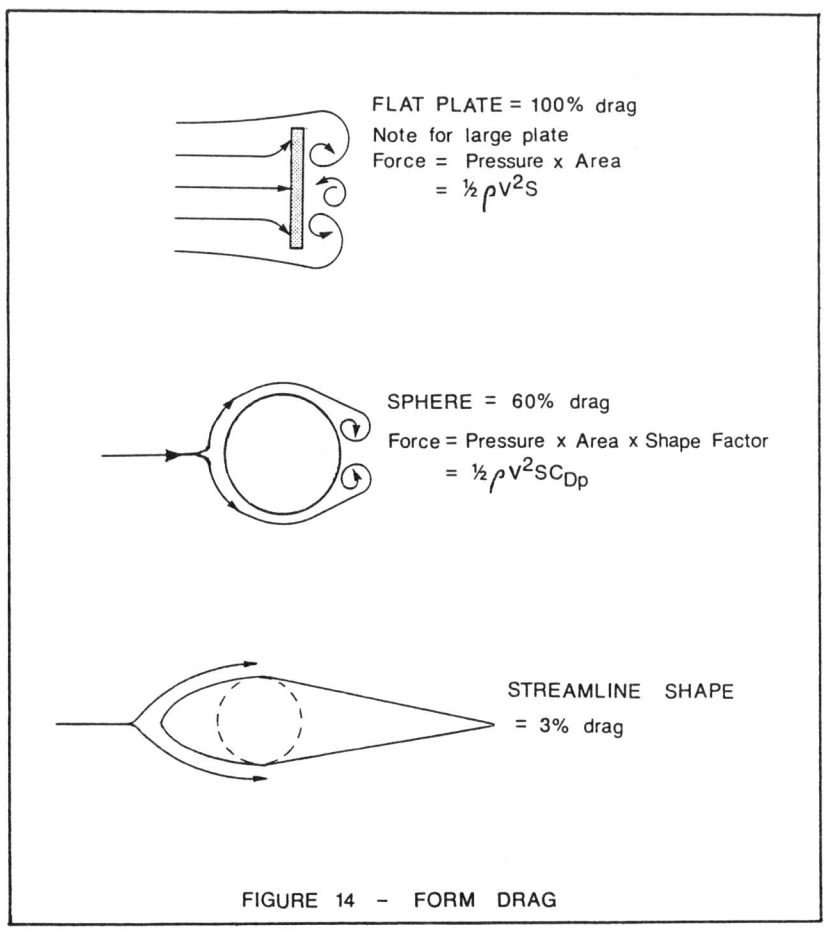

FIGURE 14 – FORM DRAG

You will probably note that the amount of drag depends to a very large extent on the amount of sideways disturbance of the air in a given time. It will have been seen that high drag means heavy turbulence; this is a reversible relationship — heavy turbulence produces high drag. Avoid the production of turbulence and you will avoid drag.

Skin Friction
When a wing or fuselage moves through the air it inevitably 'slows down' the air in contact with it. Air in contact with the wing may actually move with it, i.e. may be stationary relative to it; air a little way out will be 'slowed down' by lessening amounts as distance from the wing increases. The process means that energy must

have passed from the aircraft to air; this means, in effect, that drag has been produced. Such drag is called skin friction as illustrated in Figure 15.

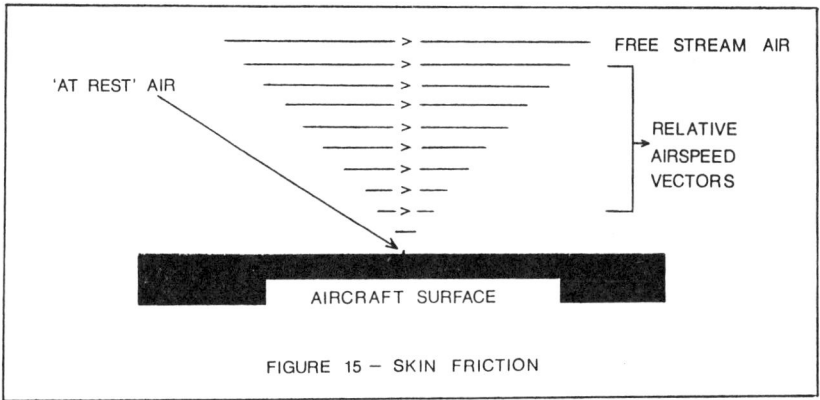

FIGURE 15 — SKIN FRICTION

Boundary Layer
The region in which the air is 'slowed down' compared with the free stream air is called the boundary layer, and may be either of two states:
   a. Laminar Flow
   b. Turbulent Flow.

Laminar Flow
Is the state of affairs where the air within the boundary layer slides across itself layer by layer without any rotational movement. Laminar flow implies a thin boundary layer and relatively low drag.

Turbulent Flow
Is the state of affairs within the boundary layer where the air has both translational *and* rotational motion; it may even flow 'backwards' along the wing. Turbulent flow implies a thick boundary layer and relatively high drag. Note that the same boundary layer may be in parts laminar and in other parts turbulent; the point where laminar flow changes to turbulent is called the *Transition Point.* In some conditions of flight the boundary layer may actually separate from the wing, leaving a region of low-energy, very disturbed air in contact with the wing. Such a condition is always associated with loss of lift and increase in drag and is known as 'flow separation'. It can occur at high and low speeds, and it is interesting to note that a turbulent boundary layer separates less easily than a laminar boundary layer as illustrated in Figure 16.

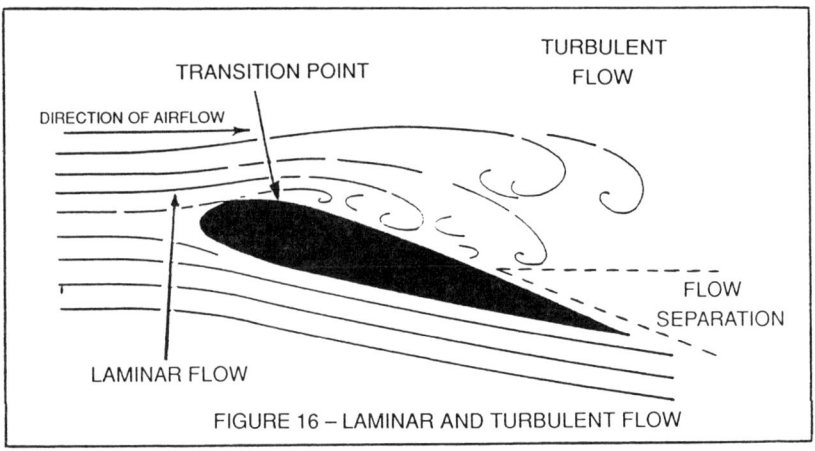

FIGURE 16 – LAMINAR AND TURBULENT FLOW

It is clear that a dominant factor in determining the thickness of the boundary layer, the nature of the flow and therefore the amount of skin friction is the state or condition of the surface of the aircraft. The smoother and more highly polished the surface, the thinner the boundary layer and the lower the drag.

Boundary Layer Control
If the thickness and condition of the boundary layer can be controlled then drag can be kept to a minimum or separation delayed. Methods of boundary layer control will be discussed later.

Induced Drag
The component of total drag which is produced because lift is generated is called induced drag and will be considered later when lift has been fully investigated.

Wave Drag
At high speeds where compressibility effects begin to produce shock waves, an additional drag appears, called wave drag. At high Mach numbers the wave drag may be the dominant proportion of the total drag.

The drag 'family tree' is now complete and is shown below.

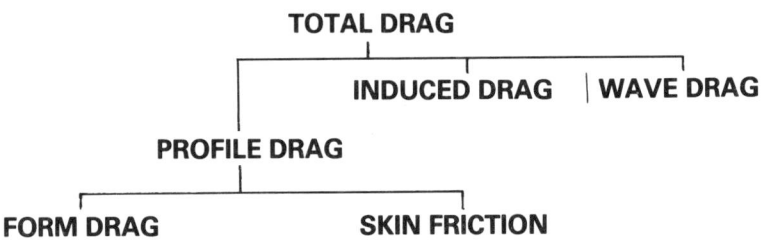

Lift

Flat Plate Lift

All that is needed to raise any object from the ground is a force greater than the weight of the object, acting in a direction opposite to the weight. The same is true for an aircraft; our job is to determine how the force is produced. Consider a flat plate inclined at a small angle to the airflow as shown in Figure 17.

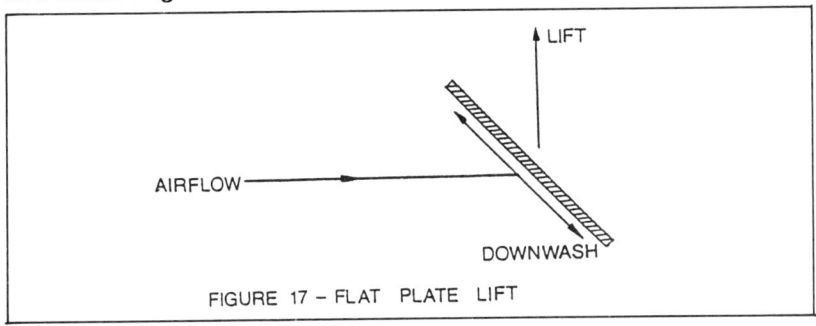

FIGURE 17 – FLAT PLATE LIFT

The main portion of the air is deflected downwards, i.e. there is a heavy 'downwash'. This must have needed a downward force on the air — by Newton's Laws, the air must have exerted an *upward* force on the plate — *Lift*.

A flat plate is, in fact, an inefficient shape because, although it produces good lift, it is structurally unsound and produces very high drag. A more efficient shape has been found, one which produces high lift with low drag and is structurally sound; such shape is called an *Aerofoil*.

Aerodynamic Language

The science of aerodynamics, like any other, has a language of its own and in order to understand the science we must be able to speak the language. This is particularly necessary in the aerofoil region and a list of definitions is given here so that no ambiguity is introduced; where two terms are permissible for the same feature, the preferred term is given first with the alternative in brackets.

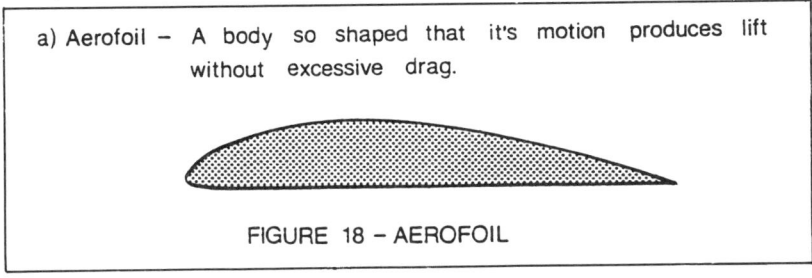

a) Aerofoil – A body so shaped that it's motion produces lift without excessive drag.

FIGURE 18 – AEROFOIL

b) Chord Line – The straight line joining the centres of the leading and trailing edges of the aerofoil section.

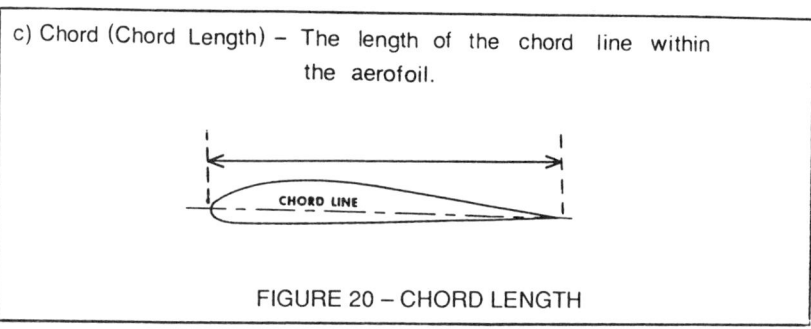

FIGURE 19 – CHORD LINE

c) Chord (Chord Length) – The length of the chord line within the aerofoil.

FIGURE 20 – CHORD LENGTH

d) Angle of Incidence (Angle of Attack) – The angle between the chord line and the direction of the free stream airflow.

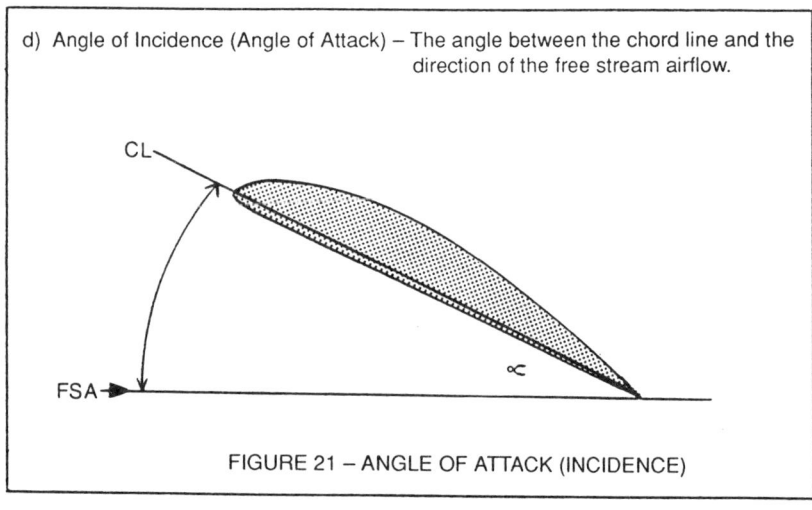

FIGURE 21 – ANGLE OF ATTACK (INCIDENCE)

e) Centre of Pressure (Aerodynamic Centre) – The point of intersection of the total air reaction vector and the chord line.

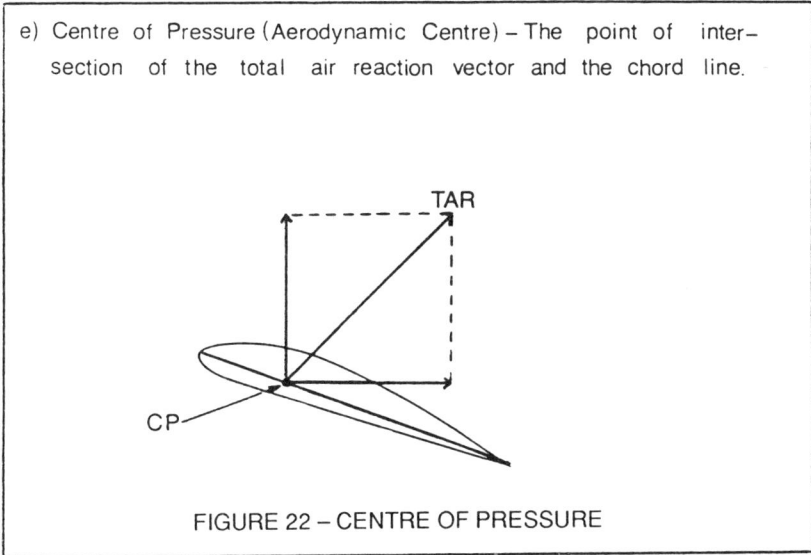

FIGURE 22 – CENTRE OF PRESSURE

The aerodynamic definitions so far discussed are shown together in Figure 23 and should be committed to memory.

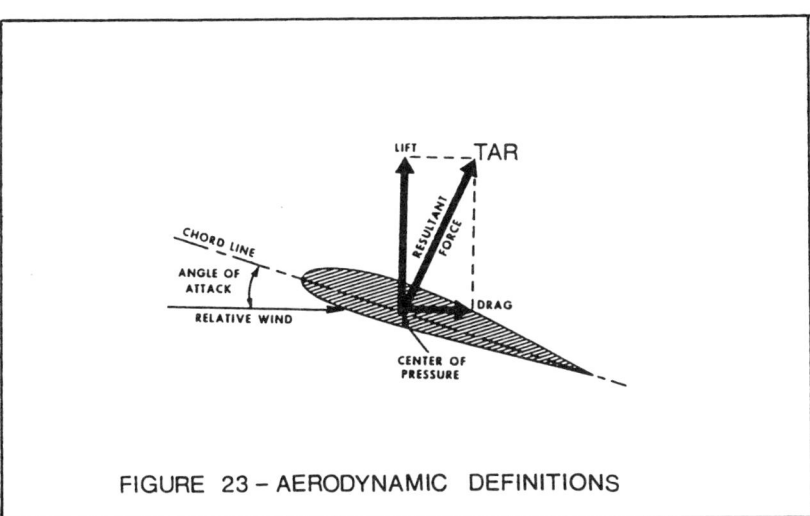

FIGURE 23 – AERODYNAMIC DEFINITIONS

Aerofoil Development

Lift can be produced by a minimum-drag streamlined shape if it is inclined at a small angle to the airflow, and by other 'cambered' shapes even at 0° angle of incidence. Consider a symmetrical streamlined section as shown in Figure 24 below.

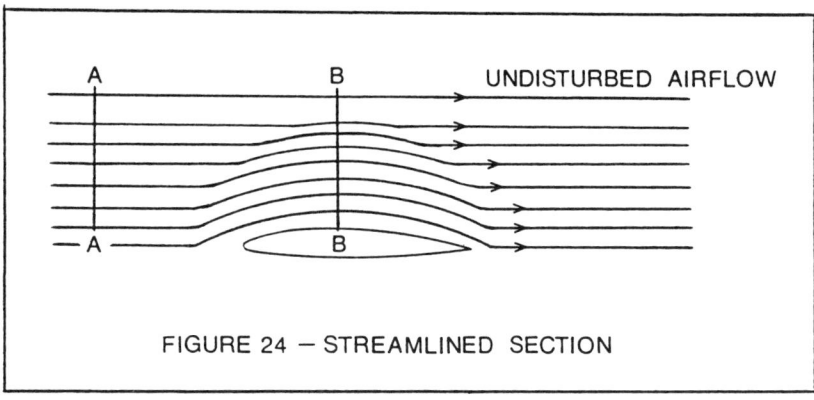

FIGURE 24 — STREAMLINED SECTION

The airflow over the section may be visualised as shown and it can be clearly seen that the air across section BB must travel more quickly than the air across section AA, because the mass flow must be the same at each point. The steady flow energy equation tells us that if speed increases, static air pressure must fall (Pa + ½ ρ V² = constant) so the static pressure at BB must be less than at AA.

With a symmetrical body at 0° angle of incidence, the static pressure reduction will be symmetrical and there will be no resultant force on the body at right angles to the airflow. The total air reaction will be parallel with the airflow, i.e. it will be entirely drag.

An asymmetric distribution will, however, produce lift as well as drag, no matter how the loss of symmetry is achieved.

a. Consider a symmetrical body inclined at a few degrees to the free stream air as illustrated in Figure 25.

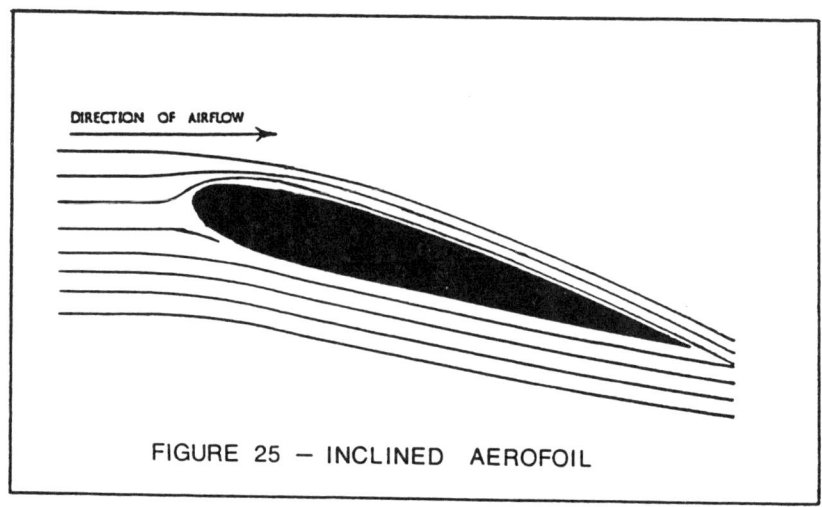

FIGURE 25 — INCLINED AEROFOIL

This state of affairs produces an overall pressure envelope which may be shown as illustrated in Figure 26.

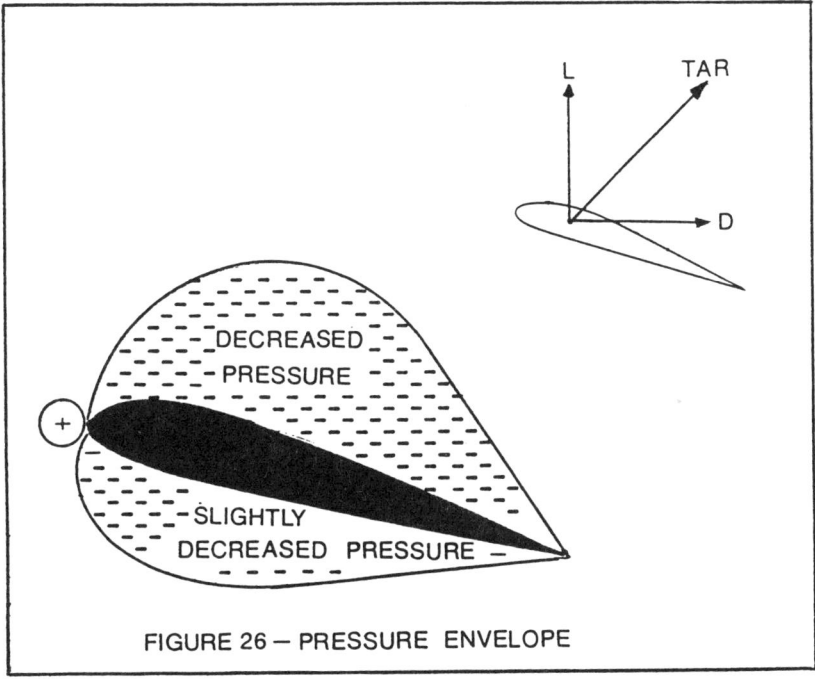

FIGURE 26 – PRESSURE ENVELOPE

Lift has been generated because of the imbalance of pressure above and below the wing as shown in Figure 26.

b. Consider now a cambered body (i.e. one whose section centre-line is curved), inclined at 0° to the free stream airflow as illustrated in Figure 27.

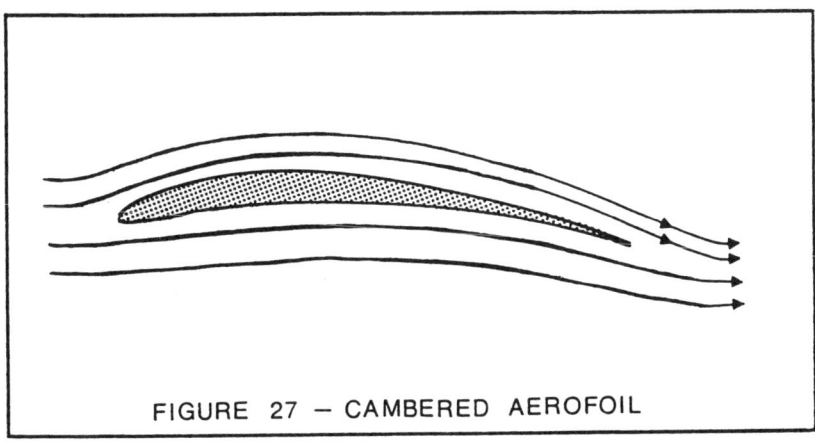

FIGURE 27 – CAMBERED AEROFOIL

Because of the camber, airflow over the upper surface must travel faster than airflow over the lower surface to maintain mass flow. The pressure reduction is therefore greater above the surface and extra lift is produced as illustrated in Figure 28.

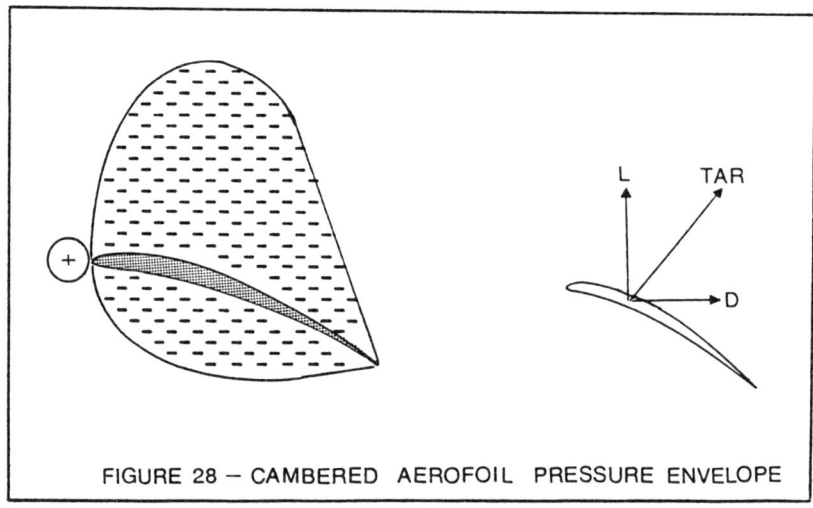

FIGURE 28 – CAMBERED AEROFOIL PRESSURE ENVELOPE

Much more lift can, of course, be produced by setting the aerofoil at positive angles of incidence. Typical pressure distribution for slightly cambered wings for variations in angles of incidence are illustrated in Figure 29.

Points of note are the presence of zero or negative lift at negative angles of attack and the possibility of appreciable pitching moments at angles where the pattern is not firmly established. Note also the growth of the lift and forward shift of the position of maximum suction as the angle of incidence increases, up to (on this section) about 16°. At around this angle, there is a sudden decrease in lift and the wing is said to have reached the 'stall'.

Approaching the stall
As angle of incidence is increased from zero degrees, downwash increases, the lift generated by the section increases and the centre of pressure moves forward (on most aerofoils). Flow chamber investigations shows that turbulence increases as angle of incidence rises, giving increasing drag above about 4° angle of incidence. At about 15° or 16° on ordinary aerofoils, flow separation becomes pronounced and the pressure distribution becomes ragged, giving a sudden reduction in lift as illustrated in Figure 30.

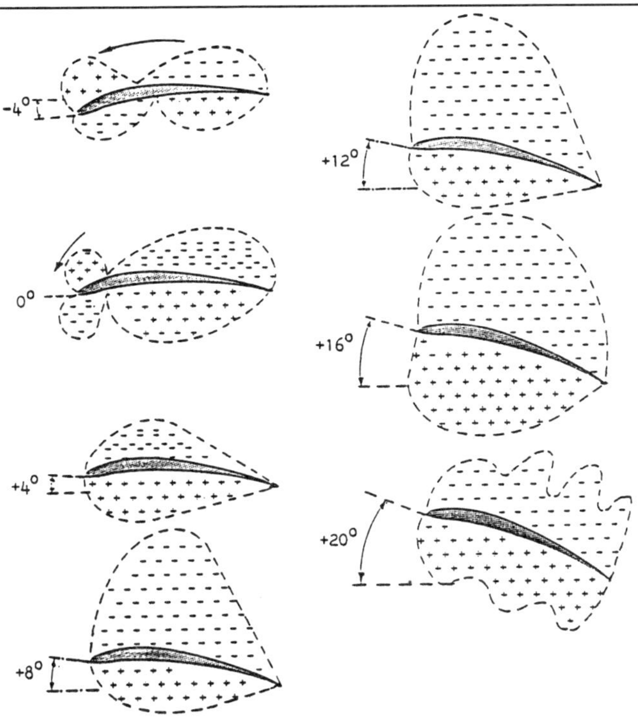

Aerodynamics for the Professional Pilot

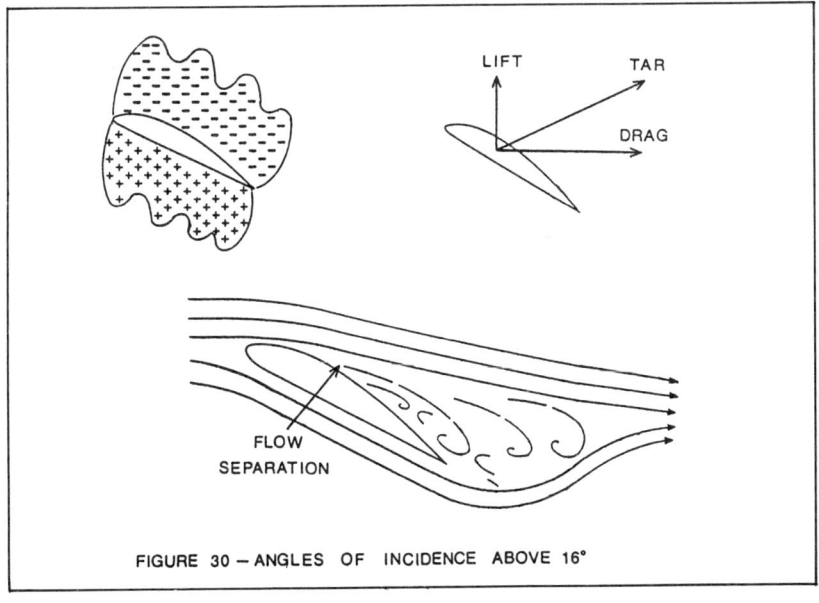

FIGURE 29 – PRESSURE DISTRIBUTION WITH VARIATIONS OF ANGLES OF INCIDENCES

FIGURE 30 – ANGLES OF INCIDENCE ABOVE 16°

This is the stall, a low lift, high drag condition, usually accompanied by a rearwards shift in the centre of pressure, a nose down pitching moment, and *always* a large loss of height in recovery. On some high performance aircraft, especially those with rear-mounted engines and high set tailplanes, the stall may be irrecoverable and therefore catastrophic.

Stagnation

Practical investigation will show that there is a point somewhere on the leading edge of the aerofoil at which some of the air is brought to rest relative to the wing. The at-rest condition is called 'stagnation' and the point at which it occurs is called the 'stagnation point'. The higher-than-ambient pressure in the stagnation region is called the stagnation pressure and as angle of incidence increases the stagnation point tends to move down the leading edge onto the lower surfaces of the aerofoil. The overall effect is to produce considerable upwash ahead of the aerofoil at large angles of incidence – a feature which can be used to operate automatic lift-augmentation devices.

Lift Equation

A drag equation was previously presented to show the relationship between drag, drag coefficient (the shape factor), air density, airspeed, and gross wing area. A similar equation may be developed for lift and turns out to be:

where
$$L = C_L \tfrac{1}{2} \rho V^2 S$$

L is the lift in lbf.
C_L is a dimensionless force coefficient called the lift coefficient and depends on the particular aerofoil section used but *also* varies with angle of incidence and other factors, such as the Mach number and aircraft configuration.
ρ is the air density in slugs/ft^3.
V is the true airspeed in ft/Sec.
S is the gross wing area in ft^2.

Thus if the quantities C_L, ρ, V and S are known or can be worked out (e.g. ρ from height) the lift may be calculated.

Induced Drag

The drag which an aircraft produces because it is generating lift is called induced drag. An explanation of its production follows, but it is only a simplified guide, whereas in actual fact to explain this action fully, a far deeper study than is possible at this stage, is required.

An increase in lift means an increase in vortex activity behind the wing. This is an indication of an increase in the total air reaction (TAR), but if we assume straight and level flight at constant weight this increase in TAR can occur only if the vector is tilted backwards, because the *Upward* force on the aeroplane remains equal to the weight as illustrated in Figure 31.

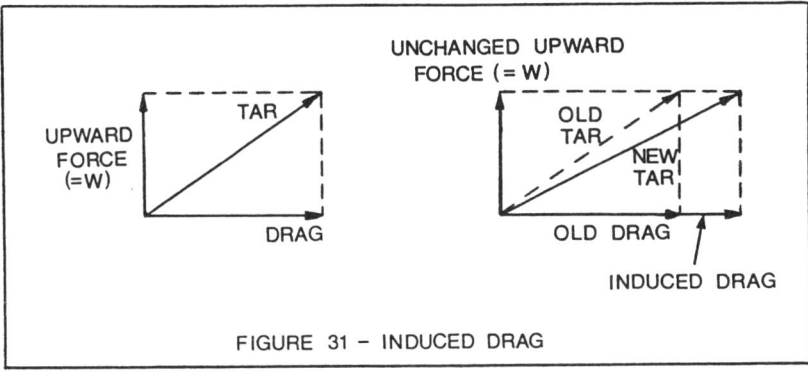

FIGURE 31 - INDUCED DRAG

The backward tilt of the TAR vector has produced an increase in the drag component which we call *Induced Drag*. To maintain the same upward force while changing the C_L a reduction in speed is necessary and it has been calculated that the induced drag is proportional to the inverse square of the true airspeed as illustrated in Figure 32.

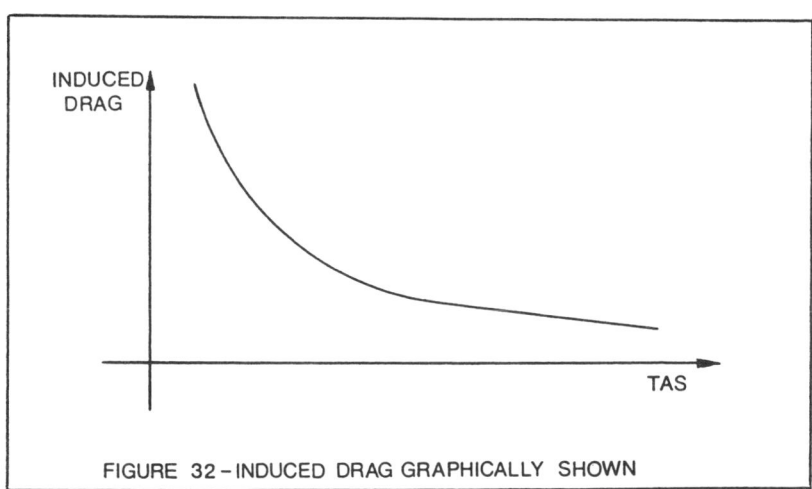

FIGURE 32 - INDUCED DRAG GRAPHICALLY SHOWN

Another way of explaining induced drag is found when we consider the energies involved. If lift is to be produced (or increased), the energy to generate the lift must be ultimately provided by the thrust from the engines as there is no other source. Since the engines operate (under steady conditions) parallel, or near parallel, to the free stream airflow, they can be said to be overcoming an extra drag. This extra drag is what we call *Induced Drag*.

Induced Drag and Aspect Ratio

A formula can be produced for induced drag similar to that for profile drag, i.e.

$$D_i = C_{Di} \; \tfrac{1}{2} \rho V^2 S$$

where ρ, V, and S are as before, and C_{Di} is the coefficient of induced drag. This coefficient is a much more complicated quantity than C_{Dp} or C_L and can be shown to be equal to

$$\frac{K C_L^2}{\pi \, AR}$$

where K is a constant which depends on wing planform as illustrated in Figure 33a and AR is the 'aspect ratio' of the wing.

$$\text{Aspect Ratio} = \frac{\text{Wingspan}^2}{\text{Wing Area}} = \frac{\text{Span}}{\text{Mean Chord}}$$

Thus

The presence of C_L in the formula should come as no surprise (we know that induced drag is dependent on lift), but a study of the full formula now shows that induced drag is *inversely proportional* to the square of the true airspeed. This is a reasonable statement when you consider that to produce a given lift, the wing has to work hardest at low airspeeds.

Tip Vortices

High induced drag is always accompanied by very active wing tip vortices (remember, turbulence = drag). Thus

The effects of these vortices are sometimes visible, especially in very wet air, as streamers behind aircraft.

Drag Coefficient

The overall drag coefficient is now given by $C_{Dp} + C_{Di}$ and the final low speed drag equation becomes:

$$D = C_D \; \tfrac{1}{2} \rho V^2 S$$

curves of drag against airspeed can now be drawn as in Figure 34.

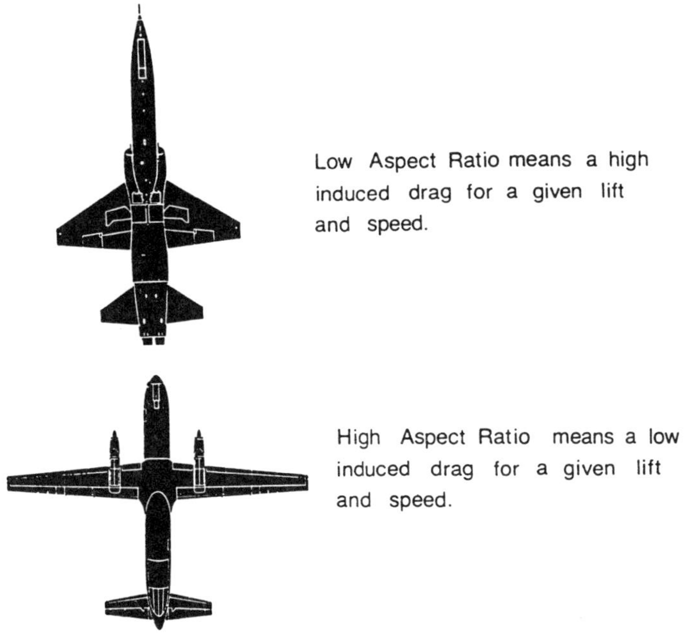

Low Aspect Ratio means a high induced drag for a given lift and speed.

High Aspect Ratio means a low induced drag for a given lift and speed.

FIGURE 33a — ASPECT RATIO

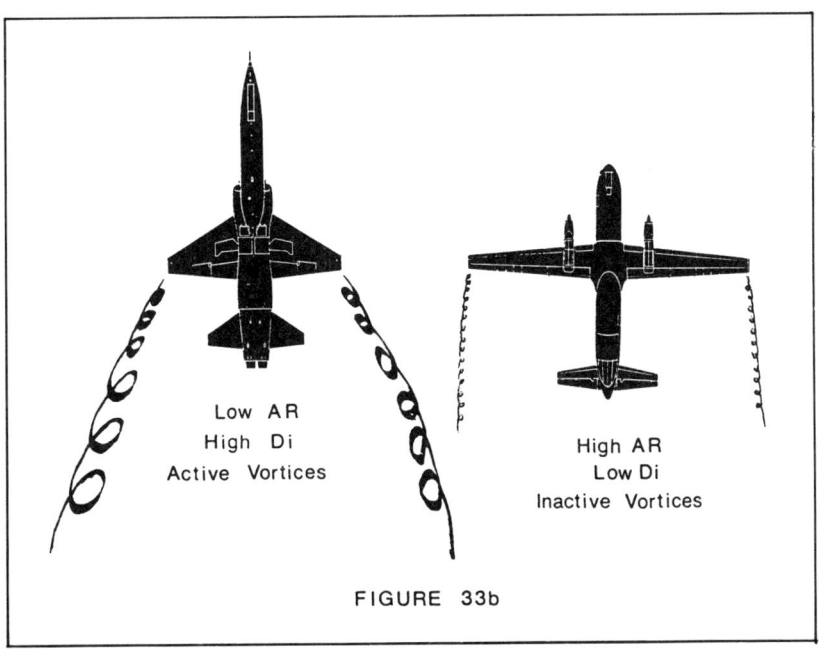

Low AR
High Di
Active Vortices

High AR
Low Di
Inactive Vortices

FIGURE 33b

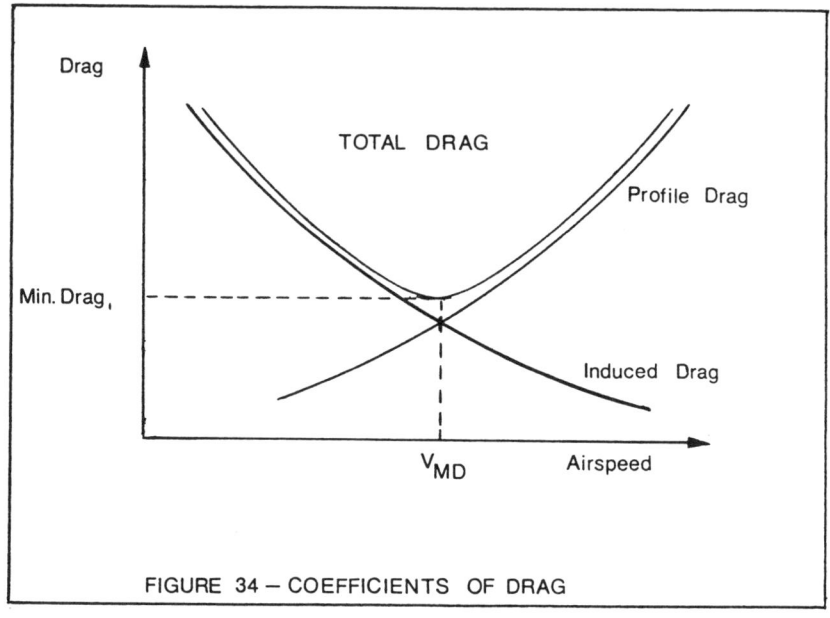

FIGURE 34 – COEFFICIENTS OF DRAG

Aerodynamic Relationships

The performance of an aeroplane depends on the relationship between the aerodynamic forces (i.e. lift and drag) and the variables in the formula $L = C_L \, \frac{1}{2} \, \rho \, V^2 \, S$ and $D = C_D \, \frac{1}{2} \, \rho \, V^2 \, S$.

For a given area and configuration, lift varies directly with dynamic pressure ($\frac{1}{2} \, \rho \, V^2$) and with the square of the airspeed. Expressed graphically we have:

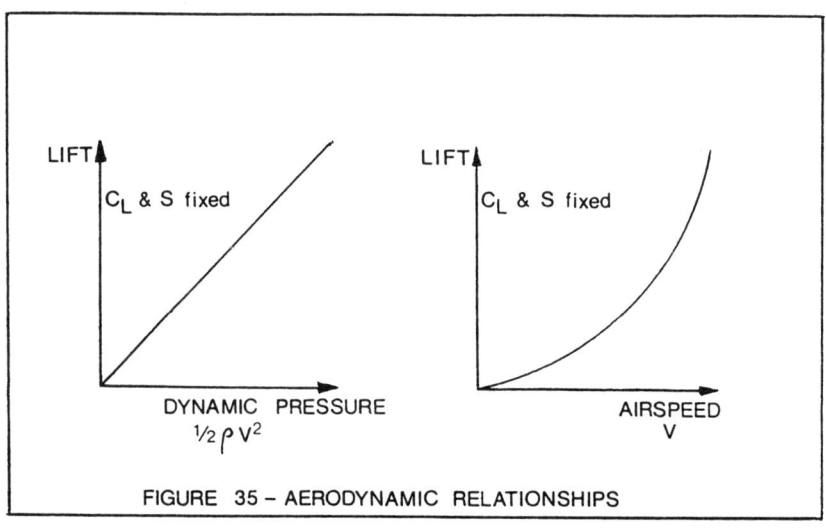

FIGURE 35 – AERODYNAMIC RELATIONSHIPS

The above mentioned graphs do not however tell the full story, because the terms which have been assumed fixed, can, and, normally do, vary. The area term, S, is usually constant but may vary with flap deflection; the C_L (or C_D) term varies with a number of factors, as we can see from the Lift equation on page 40, and the most important of these is the angle of incidence. The relationships can be shown as illustrated in Figures 36 and 37.

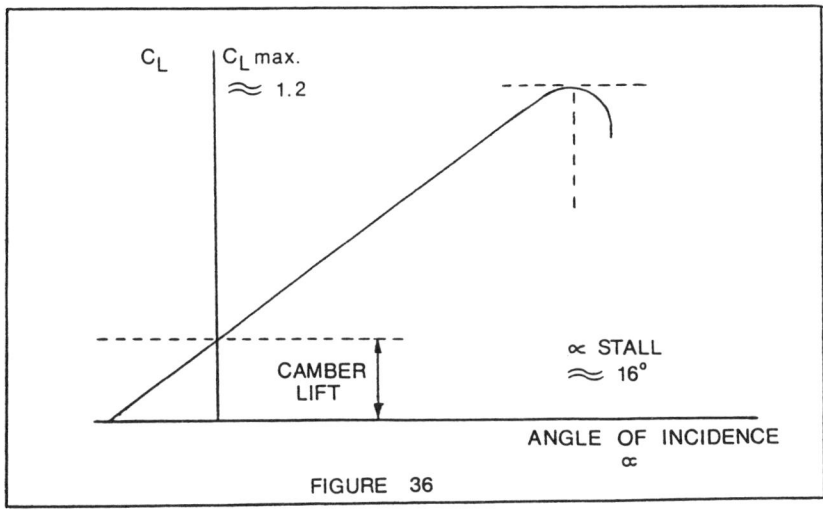

FIGURE 36

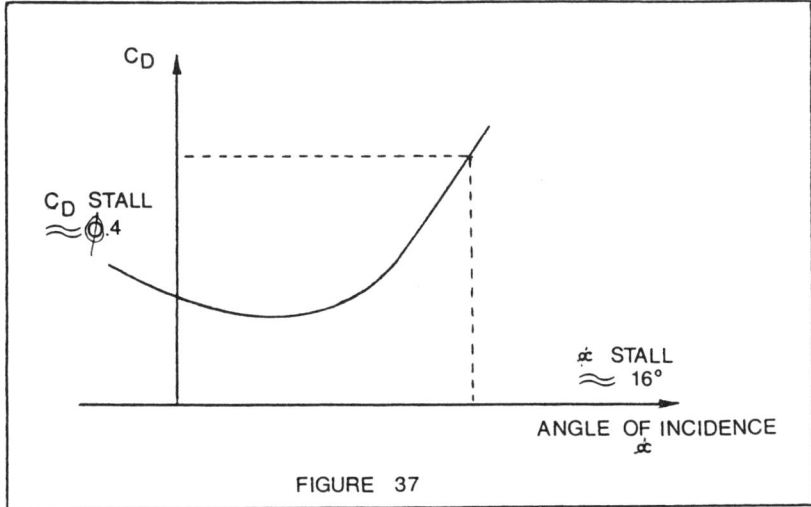

FIGURE 37

The values are typical for a conventional aerofoil and the graphs assume constant density, airspeed and wing area.

To maintain an aircraft in the air we need constant lift, assuming weight to be constant. Leaving aside the area term (as constant) we can reach any reasonable balance between C_L and airspeed; as airspeed decreases, C_L must increase, but the increase is limited at C_L max, beyond which the aerofoil stalls (as previously explained on page 38). This lowest airspeed (corresponding to C_L max) is the minimum flying speed Vmin. It follows that:

$$L = W = C_L \text{ max } \tfrac{1}{2} \rho V^2 \text{min } S$$

from which

$$V_{min} = \sqrt{\frac{2W}{C_L \text{ max } \rho S}}$$

or

$$V_{min} = \sqrt{\frac{2W}{C_L \text{ max } \rho}}$$

where W is the straight and level wing loading $= \dfrac{W}{S}$

At this stage we must think of measuring Drag against Airspeed and we are able to produce a graph to show us this as illustrated in Figure 38.

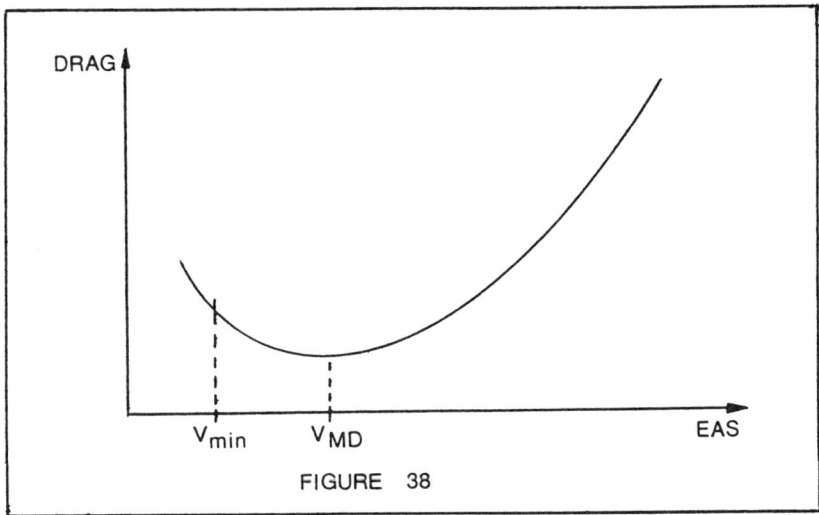

FIGURE 38

Special note should be taken of the point labelled V_{MD}, i.e. the minimum drag speed. This point is of much importance when considering the endurance of an aircraft, because it is the condition which implies minimum power settings.

Effect of Altitude
The effect of altitude on drag is best shown by plotting drag against TAS as illustrated in Figure 39.

Note: that neglecting weight changes, the minimum drag will remain the same and so will the EAS which produces it.

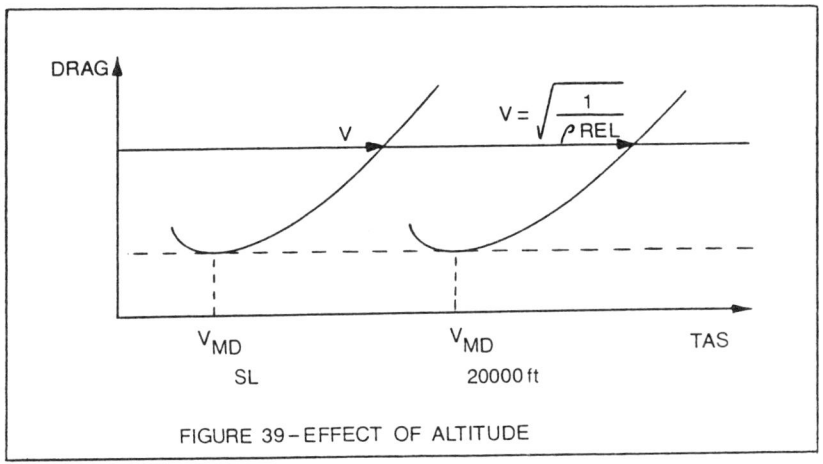

FIGURE 39 – EFFECT OF ALTITUDE

Effect of Compressibility

Above about 300 knots TAS, compressibility effects become appreciable and wave drag may cause sudden and rapid increase in drag as shown in Figure 40 following:

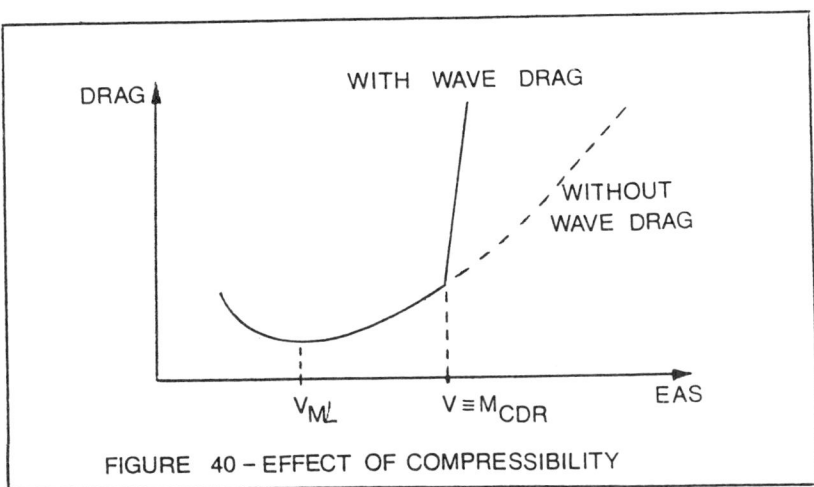

FIGURE 40 – EFFECT OF COMPRESSIBILITY

Note: that the Mach Number corresponding to the M_{CDR} point is called the 'critical drag rise' Mach Number.

Lift/Drag Ratio

A prime indication of the aerodynamic efficiency of an aeroplane is the ratio of lift to drag. A 'work-horse' type of aeroplane may have a Maximum L/D ratio of 8–12; a well designed sailplane would have a L/D ratio of perhaps 20. The L/D ratio varies with angle of incidence (because lift and drag do) and a typical relationship is shown graphically in Figure 41 below:

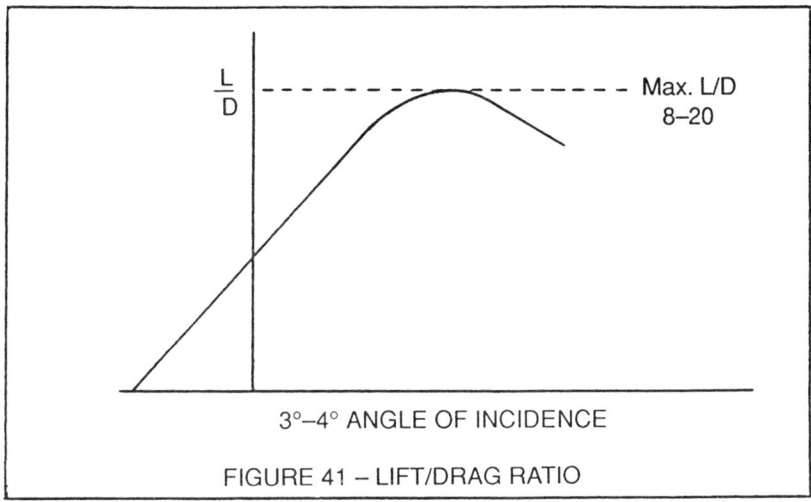

FIGURE 41 – LIFT/DRAG RATIO

There is no ideal aerofoil to suit all conditions of operation. The task of the aircraft designer is to choose the best compromise he can find, and to modify its performance, when required, by artificial means. Hence flaps, slats and lift dumpers, for landing, whereas the aeroplane cruises in a clean condition.

Boundary Layer Control
The production of lift and drag by an aircraft depends largely on the state of the boundary layer. A thin, laminar, attached boundary layer means high lift and low drag. Laminar flow may be maintained, or separation delayed, by a number of methods, such as:

    a. Flaps (leading or trailing edge)
    b. Slats/slots
    c. Fences
    d. Camber (especially leading edge)
    e. Vortex generators
    f. Sucking or blowing
    g. Notched or saw-tooth leading edges
    h. Slab trailing edges.

Some of these methods may be found on high, fast aircraft, some on low, slow aircraft. Many aircraft use a combination of two or three or more methods, e.g. slats and flaps. The basic advantage of boundary layer control may be summed up as:

a. Ability to control and increase C_{Lmax}, giving control and reduction of the minimum flying speed.
b. Control and reduction of drag.
c. Greater flexibility of structural design, e.g. with BLC, thicker wing sections may be used giving greater storage space and better integrity.

We shall restrict our detailed study of boundary layer control to the operation and effects of slats and flaps.

Slats

A slat is an auxiliary aerofoil section placed close to the upper leading edge of the wing, thus creating a narrow slot between itself and the wing. Air passing through the slot is speeded up, or energised, by the venturi effect and this energetic air is injected into what would otherwise be a turbulent or even separated boundary layer as illustrated in Figure 42.

The slat may be a fixture (giving a permanently open slot) or may be operated either by the pilot when required, or automatically by upwash at high angles of incidence.

Permanently open slots are so arranged that the slat contributes little drag at low angles of incidence; they are rare on modern aircraft. Pilot operated or automatic slats are much more usual and are currently enjoying a new lease of life on the 'second generation' jet aeroplane, e.g. Boeing 747, 737 and Tristar, etc.

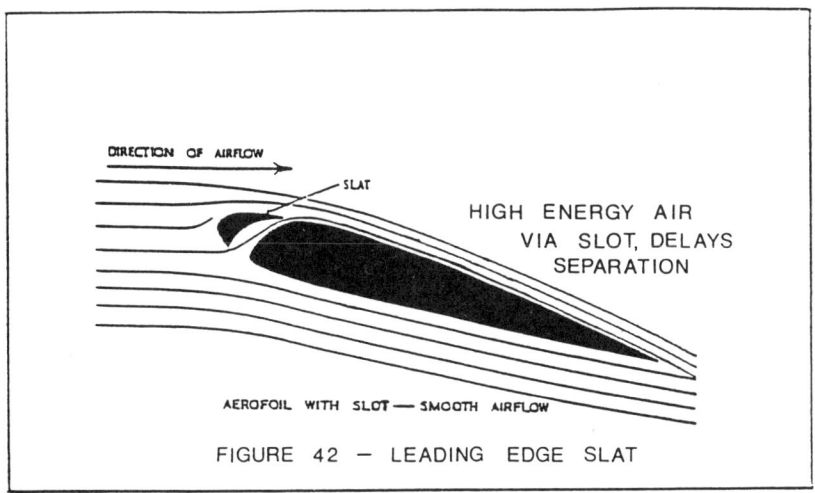

FIGURE 42 — LEADING EDGE SLAT

Effects of Slats on Aerofoil Performance

A slat/slot system considerably modifies the aerodynamic performance of its parent aerofoil. Considerably increased values of C_L and angle of incidence are obtainable and the drag rise associated with large angles of incidence is much delayed. Typical effects are summarized in the graphs in Figure 43, below:

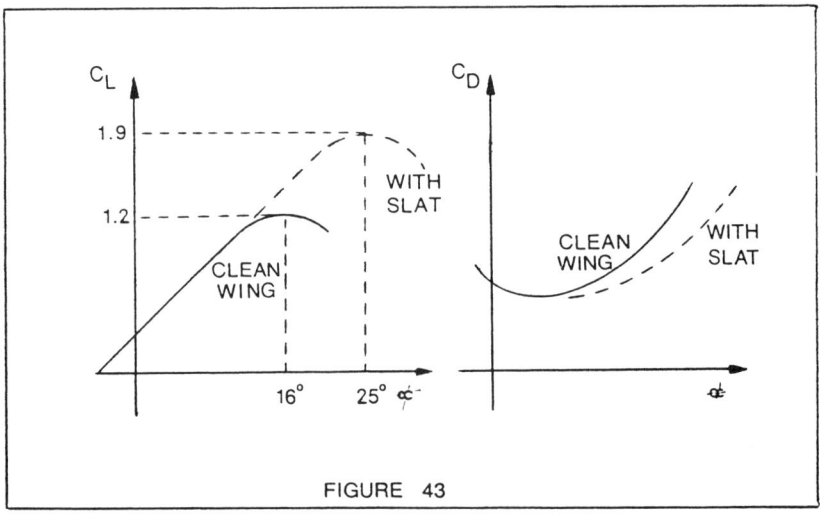

FIGURE 43

Slats alone are very seldom used on modern aircraft (the Tiger Moth was about the last design to feature slats only), because they allow the possibility of very high angles of incidence and attitude. This is not only uncomfortable and disturbing for passengers and crew, but may be extremely dangerous in the case of swept-wing high performance aircraft because it may lead to an irrecoverable 'super-stalled' condition. On some modern jet transports, the slats cannot be extended until and unless trailing edge flaps are operated; this counters the nose-up effects of the slats.

Trailing Edge Flaps

A trailing edge flap in essence is a method of changing the camber of the aerofoil. There are many varieties of flap, ranging from a simple so called plain flap, in which only camber is changed, to the triple-slotted FOWLER flap to the Boeing 727, which changes camber and wing area, and introduces slots between flap sections as illustrated in Figure 44.

The operation of a trailing edge flap alters the airflow over the whole of the wing section (and, indeed, in front of it) and the overall effect is the generation of higher C_L for any given angle of incidence. Flaps may be viewed as simply increasing downwash, thereby increasing lift; the larger the flap extension, the greater the increase in lift.

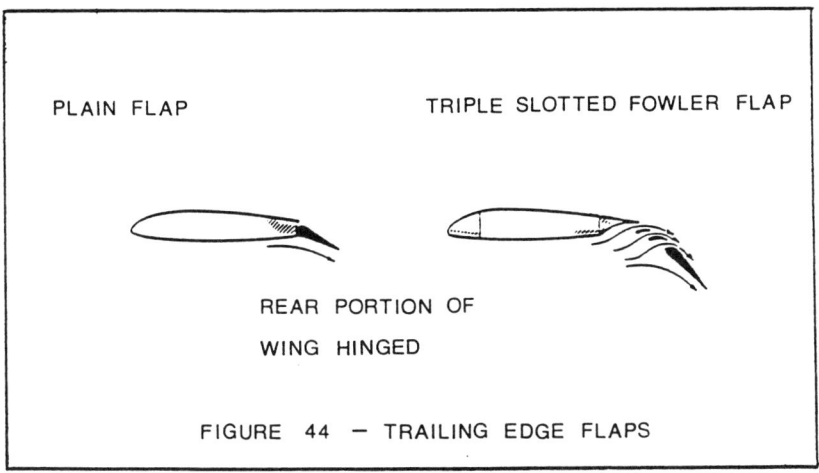

FIGURE 44 — TRAILING EDGE FLAPS

Effects of Flaps on Aerofoil Performance

We have seen already that trailing edge flaps increase the lift at a given angle of incidence. Further study shows that the maximum C_L occurs at a lesser angle of incidence than on the clean wing. Graphs of C_L and C_D against angle of incidence show the effects in Figures 45 and 46.

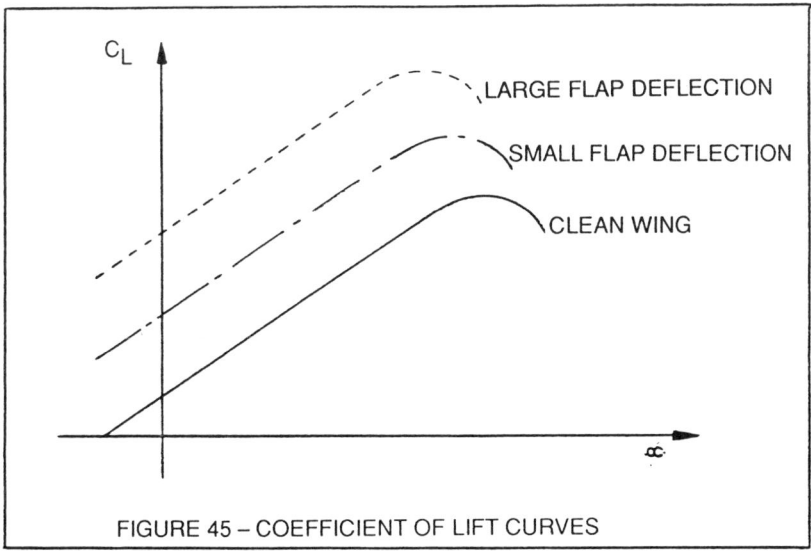

FIGURE 45 — COEFFICIENT OF LIFT CURVES

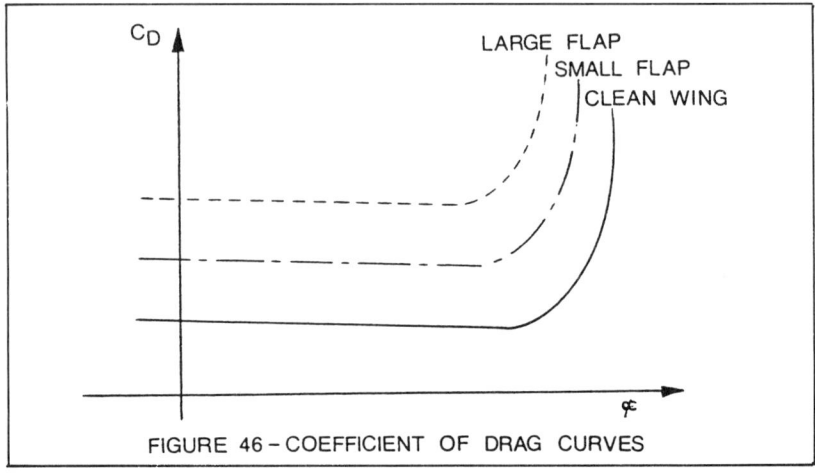

FIGURE 46 – COEFFICIENT OF DRAG CURVES

The graphs show the increase of C_L and corresponding decrease in stalling angle. They show also the increase in drag caused by flap deflection; for small deflections, the drag rise is swamped by the increase in lift, but for large deflections the increase in drag is proportionately much greater than the increase in lift. Thus for small deflections the L/D ratio increases, but for large deflections the L/D ratio decreases. These relationships have important implications when take-off and landing cases are considered.

Jet Flaps and Blown Flaps

Some research has been done into the possibility of ejecting curtains of fast moving air from wing trailing edges, thus producing jet flaps as illustrated in Figure 47.

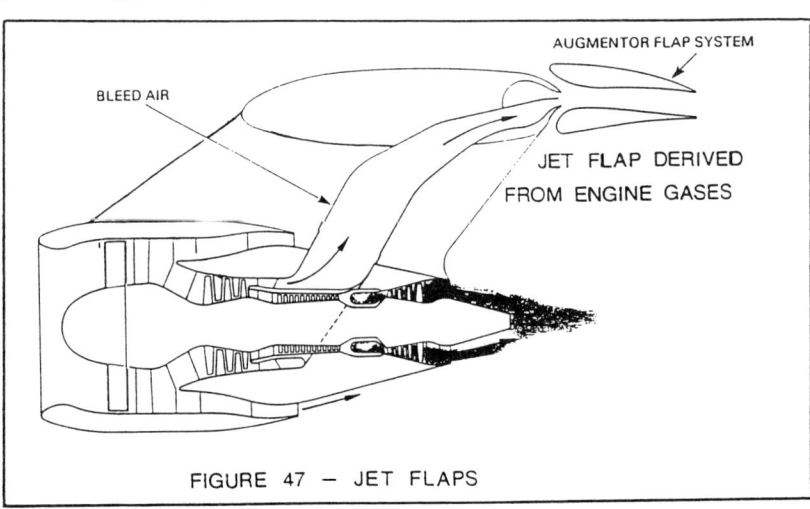

FIGURE 47 — JET FLAPS

There is some advantage in this method (e.g. no moving parts) but it imposes a considerable demand on compressor or turbine air and may well therefore drain power from the engines when it is most needed. The method has not been used on any production aircraft so far. Much more use has been made of the blown flap, in which air is blown over the flap in order to control the boundary layer.

Such a system is used on the Buccaneer, which features air blowing over leading edge, flaps, aileron and tailplane as illustrated in Figure 48.

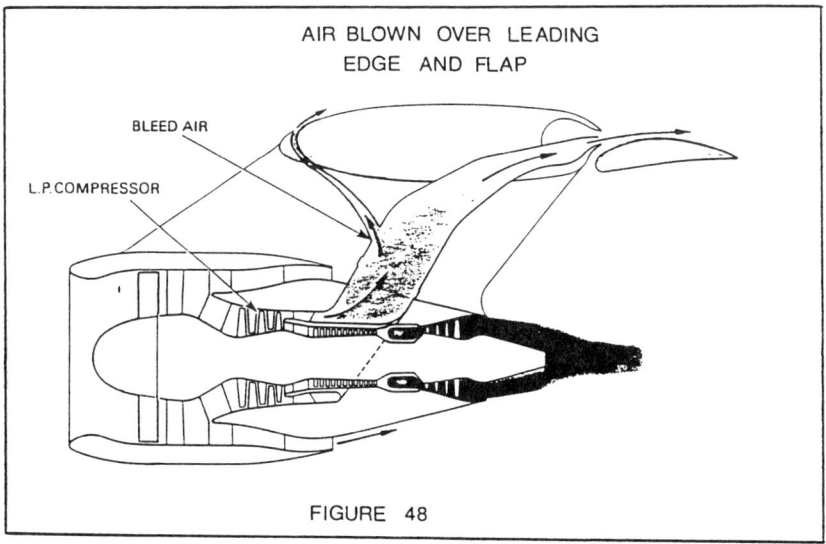

FIGURE 48

5.
Control and Stability

Aircraft Axes – Control – Stability – Centres of Gravity and Pressure – Design Effects

5 CONTROL AND STABILITY

Aircraft Axes

An aeroplane manoeuvres in flight about its centre of gravity. In order to explain the various methods of control and the achievement of the required degree of stability, it is first necessary to define the system of aircraft axes.

Three mutually perpendicular axes are imagined to pass through the C of G as illustrated in Figure 49.

   a. The Longitudinal axis (nose to tail), about which the aircraft rolls.

   b. The lateral axis (wing tip to wing tip), about which the aircraft pitches.

   c. The normal axis, about which the aircraft yaws.

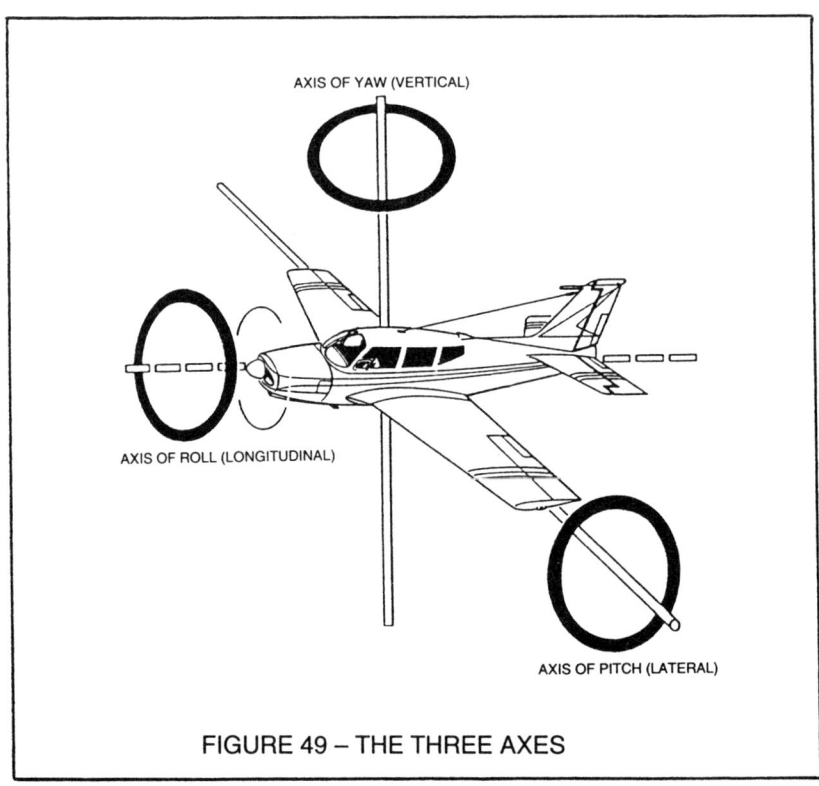

FIGURE 49 – THE THREE AXES

Control

Control Systems

The 'conventional' flying control system uses an elevator – aileron – rudder combination. The required change of attitude is achieved by changing the lift on the relevant part of the airframe.

Suppose for instance that a nose down pitching movement is required. The trailing edge of the tailplane is depressed and this produces a larger effective camber and a larger effective angle of incidence on the tailplane, which gives extra tailplane lift and therefore an upward movement of the tail as illustrated in Figure 50.

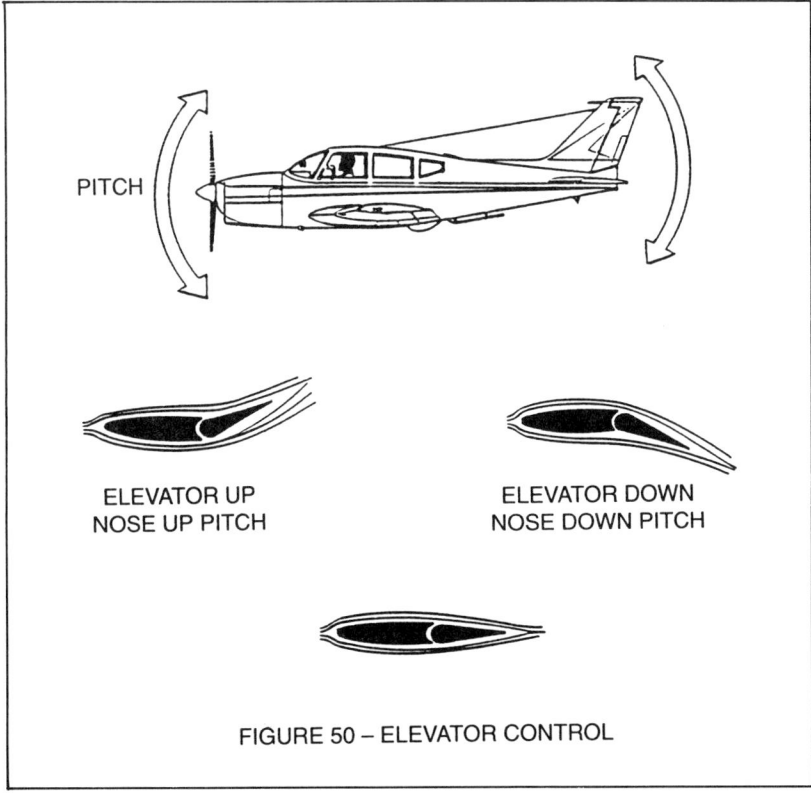

FIGURE 50 – ELEVATOR CONTROL

An upward movement of the elevator produces the opposite effect and results in a nose up movement. Exactly the same procedure is followed about the other two axes using the ailerons (rolling) and the rudder (yawing). In the case of the ailerons, one is depressed as the other is raised, producing increased lift on one wing and decreased lift on the other, thus rolling the aeroplane as seen in Figure 51.

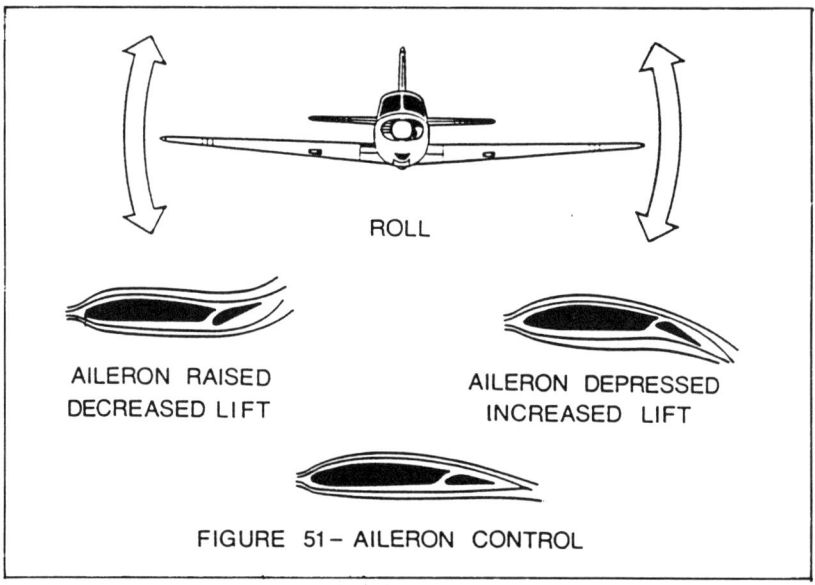

FIGURE 51 – AILERON CONTROL

The fin and rudder combination does the job in the yawing plane. Deflection of the rudder to the left, for instance, produces lift on the fin as shown below in Figure 52, and the aeroplane yaws to the left.

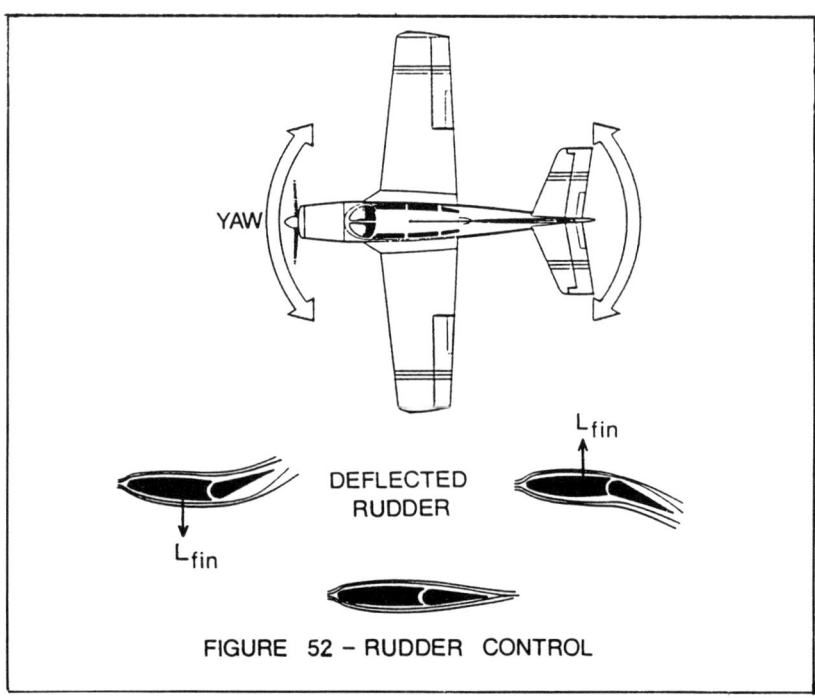

FIGURE 52 – RUDDER CONTROL

The fin is usually a symmetrical aerofoil section and it is worth noting that a fin-stall is inevitably catastrophic – the designer, in fact, limits the rudder power so that a fin stall cannot happen.

Many types of control systems have been used, either in addition to or instead of the 'conventional' system. Examples are given below but in all cases their ultimate actions are the same, they change the lift of the relevant portion of the airframe. Such systems may be:

a. All flying tailplanes (moving tailplane with elevators)
b. Slab tailplanes (one piece)
c. All moving fins (one piece)
d. Elevons (combined ailerons and elevators)
e. Tailerons (differentially moving tailplane)
f. Air jets (required on the vertical take-off aircraft)
g. Spoilers (decrease the lift on one or both wings).

Adverse Yaw

A particular design problem arises when ailerons are operated. If the ailerons are deflected by the same angular amount (one up, one down), the down going aileron creates more drag than the upgoing one (remember the reasons for induced drag). The increased drag on the upgoing wing yaws the aeroplane towards that wing, and this yaw must be met by 'in-turn' rudder. The designer minimizes the effect by using such techniques as differential aileron movement (in which the downgoing aileron is deflected *less* than the upgoing one) or by using *Frise* ailerons in which the upgoing aileron projects a beak into the airflow, thus increasing its drag as illustrated in Figure 53.

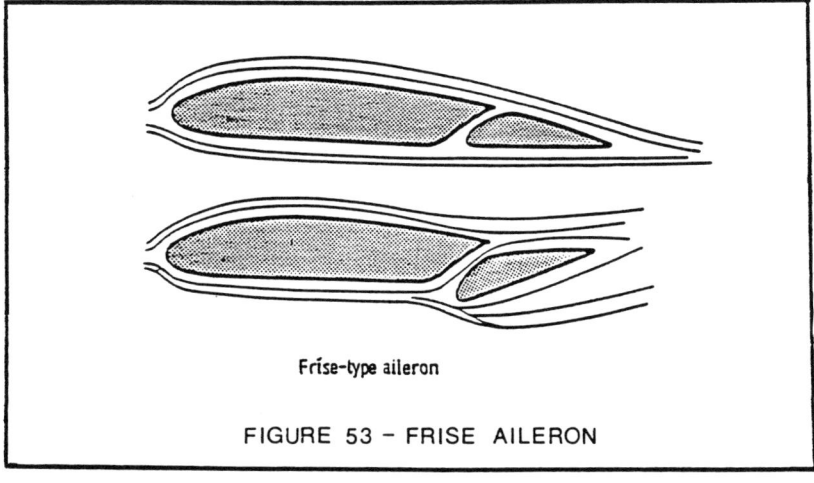

Frise-type aileron

FIGURE 53 – FRISE AILERON

Other attempts to equalize the drag may use a combination of slots and aileron, or of spoilers and aileron.

Aileron Reversal
A phenomenon known as aileron reversal may occur under certain aerodynamic conditions:

a. At very low speeds, associated with high angles of incidence, downward deflection of an aileron may stall the wing, or at least reduce the lift on it. The effect is then to cause the wing to drop rather than to rise – a reversal of the aileron control. For this reason, at low speeds a wing is 'picked up' by use of rudder, not by the ailerons
b. At high EASs, the deflection of an aileron may produce such large torsional effects on the wing that the whole wing section twists. Thus downgoing aileron will twist the wing downwards, decreasing the angle of incidence rather than increasing it; this decreases the lift instead of increasing it and the wing drops. The designer must make sure that the torsional stiffness of the wing is such that it will not twist sufficiently to produce aileron reversal, or that the speed at which it will occur is beyond the design diving speed of the aircraft.

Stability

Introduction
An aeroplane is said to be stable if it tends to return to its original trimmed condition after having been displaced. It should be noted that we are concerned only with the inherent stability of the aircraft, not with the pilot's response to displacement; in other words, in all the work which follows we shall assume that the pilot does nothing, and any response from the aircraft arises purely because of its design features.

Static Stability
An aeroplane is statically stable, if, when disturbed from straight and level flight, it tends to return to the equilibrium position. If it tends to depart further from the equilibrium position it is statically unstable. If it remains in the position to which it was disturbed it is said to have neutral static stability.

Dynamic Stability
All aeroplanes are statically stable, or at least, neutrally statically

stable, but the possession of static stability is only part of the stability story. An aeroplane may tend to regain its original trimmed state but may overshoot it and diverge from it; or it may set up an oscillatory motion called a *Phugoid,* which may or may not be convergent. Thus we have several conditions of dynamic stability of an aeroplane, as shown in the following Figures:

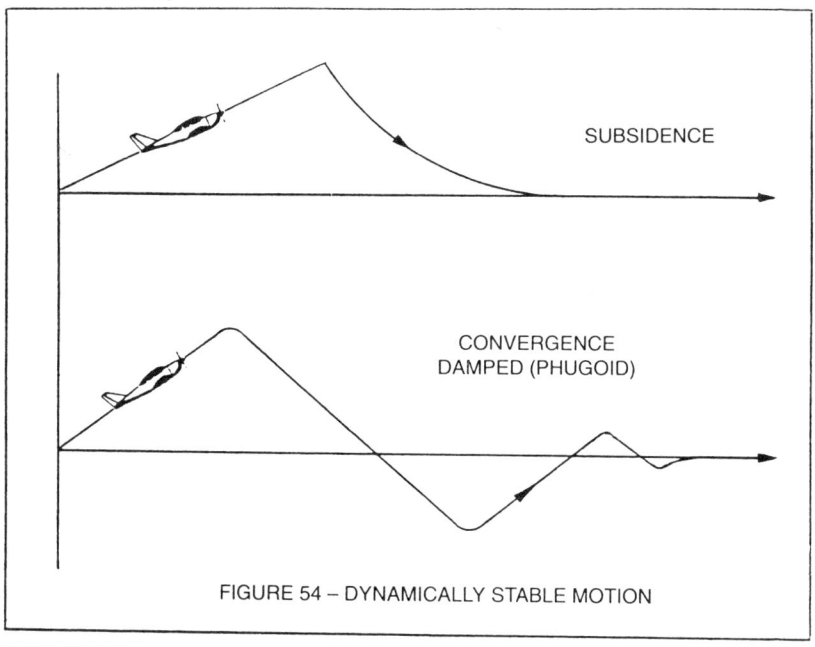

FIGURE 54 – DYNAMICALLY STABLE MOTION

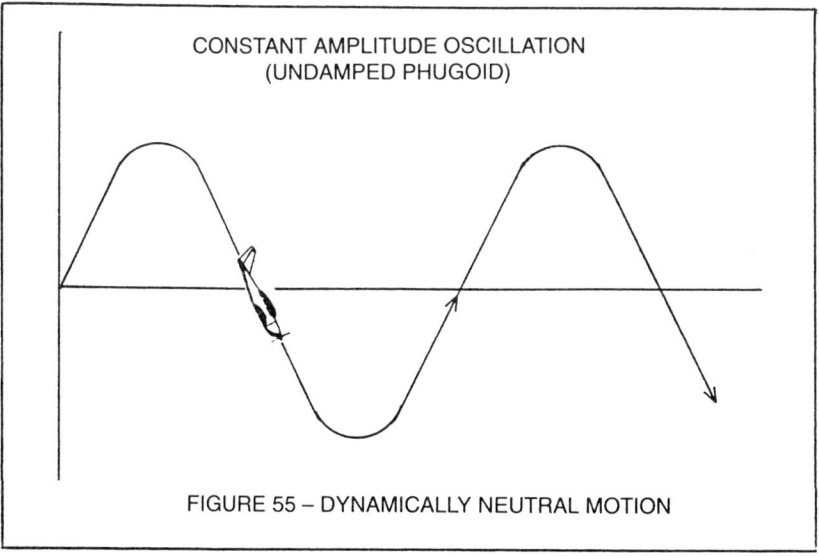

FIGURE 55 – DYNAMICALLY NEUTRAL MOTION

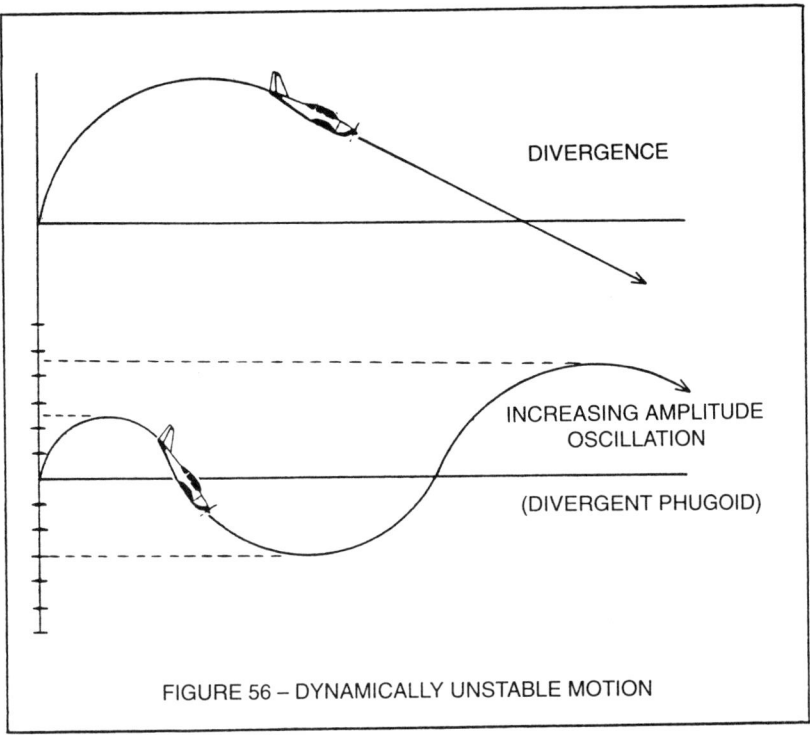

FIGURE 56 – DYNAMICALLY UNSTABLE MOTION

It will be seen that if an aeroplane is statically unstable it must be dynamically unstable. If an aeroplane is statically stable it may or may not be dynamically stable. Some aeroplanes may need artificial aids (yaw dampers, mach trimmers, etc) to achieve the desired degree of stability. Some aeroplanes may be stable about one or two axes but unstable about others. The degree of stability may vary with EAS, altitude, mach number and other factors.

Longitudinal Static Stability
The longitudinal static stability of an aircraft, which governs its level flight characteristics, is influenced by three main factors:

a. Position of the centre of gravity
b. Position of the centre of pressure
c. Tailplane and elevator design.

Position of CG
An aeroplane is most stable when its CG is furthest forward. If trimmed with a forward CG, then displaced, the aircraft will quickly return to its trimmed condition. Heavy stick forces are necessary to manoeuvre the aeroplane and it may become uncontrollably nose

heavy at low airspeeds. This is especially important in the landing case when full elevator may not be sufficient to raise the nose for the round out. As the CG is moved aft, the aircraft becomes less and less statically stable, until a point is reached where it remains in the disturbed condition; neutral stability has been reached. This may well represent the aft limit of the CG, because any further backwards movement produces instability and divergence.

Position of CP
The position of the CP is important because unless the CP coincides with the CG, pitching moments are set up. The CP, in fact, moves considerably in flight, being a function of the angle of incidence, which itself varies with:

- a. Weight
- b. Normal acceleration
- c. Turbulence
- d. Configuration.

Tailplane Design
The function of the tailplane is to counter any undesirable pitching moments set up by the aerodynamic forces. The tailplane is so designed and rigged as to produce greater restoring moments than the original disturbing moments. For example, suppose that the CP changes position because of increase in angle of incidence. Wing lift increases and the distance of CP from CG changes giving an overall pitching moment equal to:

$$\text{Lift increase} \times \text{distance of CP from CG}$$

It is the job of the designer to make sure that the associated change in tailplane lift (remember the tailplane has increased angle of incidence too) multiplied by its distance from CG is greater than the wing lift moment.

Thus the interaction between CG, CP and the tailplane determines the degree of longitudinal static stability and the sum of the effects of the three factors decides the degree of stability in steady, level flight.

Longitudinal Dynamic Stability
Stability in manoeuvring (i.e. accelerated) flight is an extremely complicated study and we shall confine ourselves to saying that the exact amount of stability built into an aeroplane is decided by the designer and it varies with the role. Large aeroplanes are usually very stable; small, fighter aeroplanes are usually much less stable because high manoeuvrability is required.

Directional Stability

The degree of directional stability possessed by an aeroplane depends on the relative effects of the keel surfaces in front of and behind the CG. It is the job of the fin to provide directional stability; without a fin most aircraft would be directionally unstable because of the CP of a pear-drop shaped body (a typical fuselage shape) is ahead of its CG. The fin provides 'weathercock' stability, as illustrated in Figure 57.

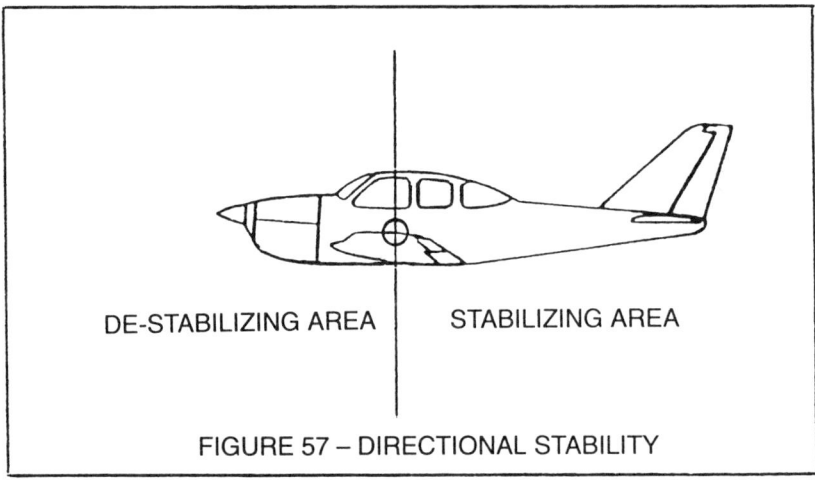

FIGURE 57 – DIRECTIONAL STABILITY

Lateral Stability

When an aeroplane is disturbed laterally, the initial rolling motion produces a larger angle of incidence on the downgoing wing than on the upgoing wing as illustrated in Figure 58.

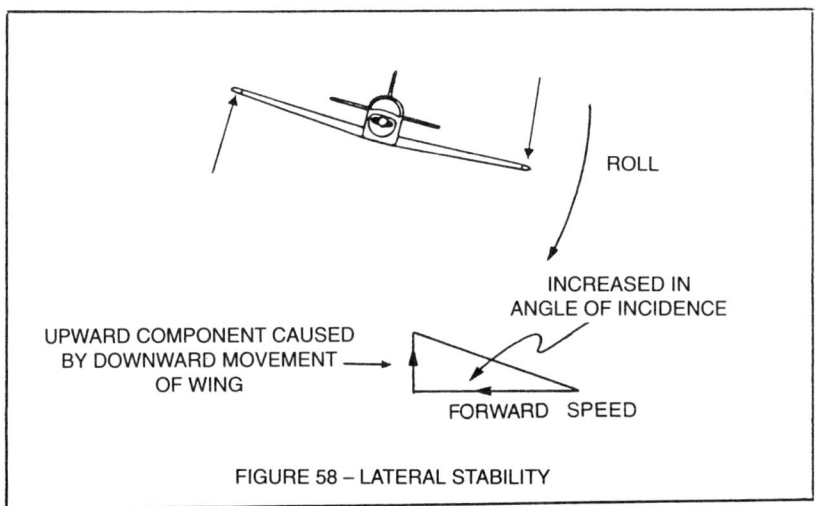

FIGURE 58 – LATERAL STABILITY

This means that *whilst* the aeroplane is rolling, the imbalanced lift provides a restoring moment which may well stop the roll, but which disappears when the roll stops. The inclination of the lift vector in the banked condition produces a horizontal component which causes the aircraft to sideslip, resulting in side loads on the aircraft. If these side loads tend to restore the aircraft to its original undisturbed state it is said to be laterally stable. Methods used to ensure lateral stability are:

a. Inclining wings (dihederal effect)
b. Sweepback
c. High keel surface relative to CG
d. High wing and low CG combination.

Dihedral Effect
Consider an aeroplane with its wings inclined as shown below in Figure 59:

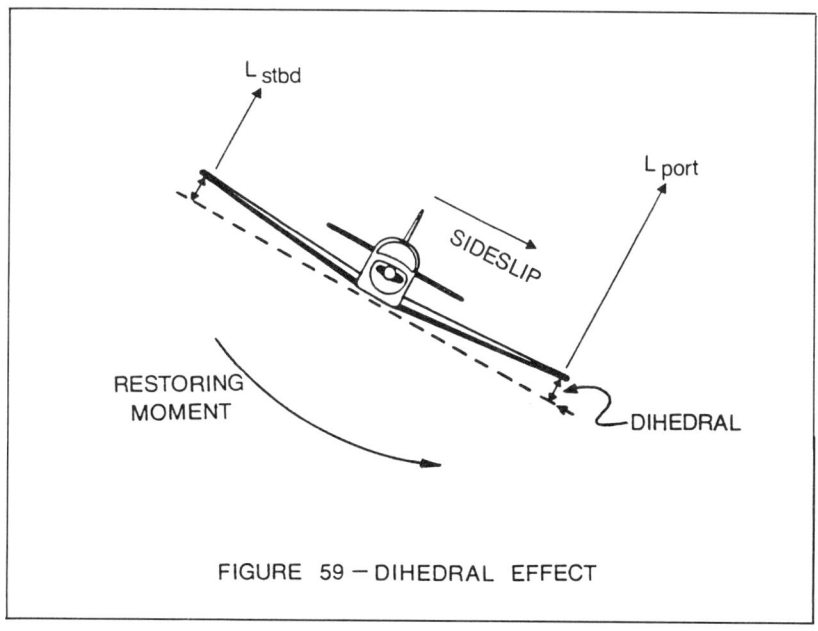

FIGURE 59 – DIHEDRAL EFFECT

The airflow meets the lower wing at a higher angle of incidence than it meets the upper wing and the lower wing therefore produces more lift than the upper wing. The imbalance produces a moment which restores the aircraft to its original state. Such wing inclination is called *Dihedral*; the reverse condition, i.e. wings inclined downwards is called *Anhedral* as illustrated in Figure 60, and produces lateral instability, however it produces manoeuvrability.

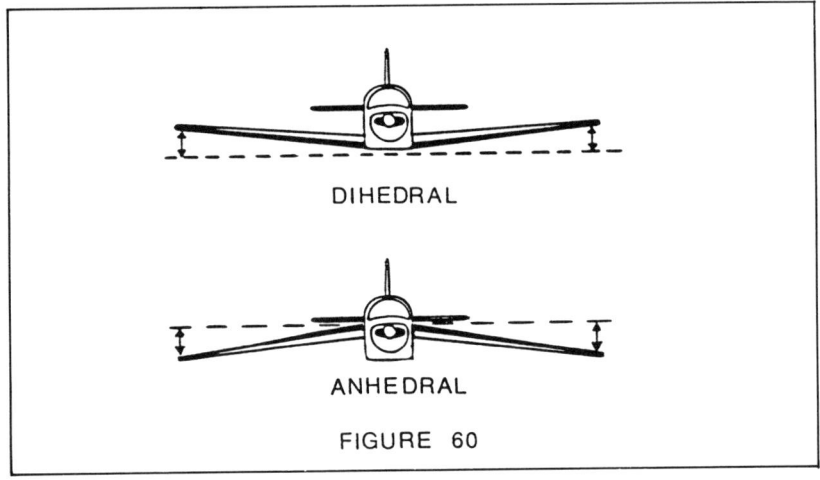

DIHEDRAL

ANHEDRAL

FIGURE 60

This stabilizing effect of dihedral is assisted by the shielding of the upper wing by the fuselage.

Sweepback

When an aircraft with swept wings sideslips the lower wing represents a shorter effective chord and therefore a greater effective camber to the airflow than the upper one, as illustrated in Figure 61.

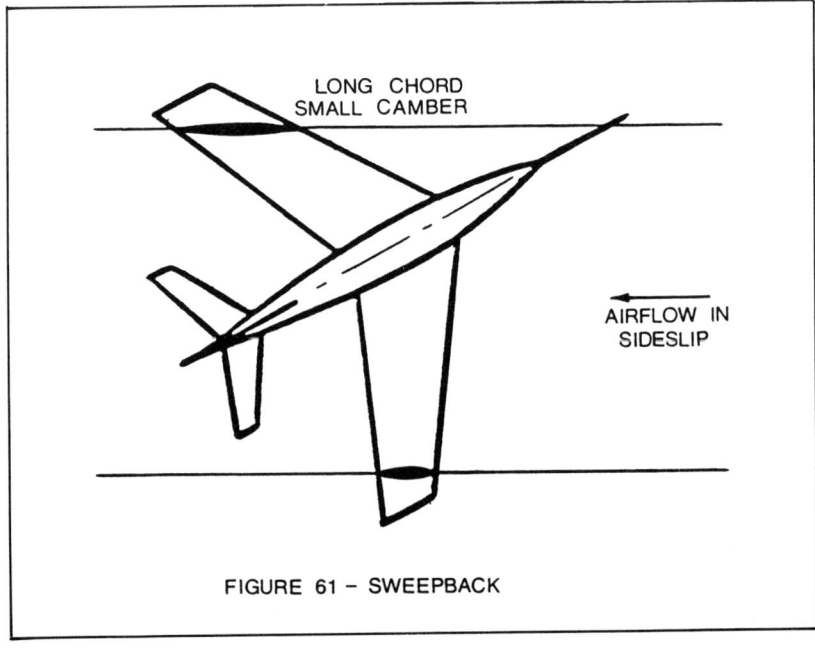

FIGURE 61 – SWEEPBACK

The effect is to produce greater lift on the lower wing than on the upper, resulting in a restoring moment. The effect is again assisted by the shielding of the upper wing by the fuselage. Aircraft with heavily swept wings are often far too stable and their wings are anhedralled to remove the unwanted stability.

High Keel Surface

If much of the keel surface of an aircraft is above the CG the side loads produced in a sideslip will tend to right the aeroplane, as illustrated in Figure 62, thus contributing to lateral stability.

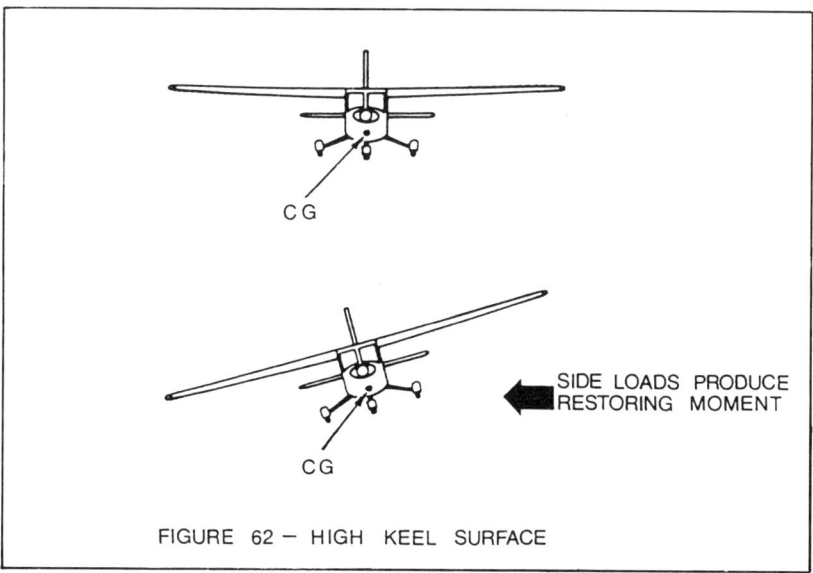

FIGURE 62 — HIGH KEEL SURFACE

High Wing and Low CG

The effect of a high wing and low CG combination is known as pendulum stability. In the sideslip which results from a lateral disturbance, the drag of the wing allows the CG to swing down, thus restoring the aircraft to its original state.

Interaction of Lateral and Directional Effects

The disturbance of an aeroplane in either lateral or directional sense always involves the other plane when the aeroplane reacts. For instance a simple yaw produces roll because the outer wing moves faster than the inner one and thus produces more lift. There are two main combinations of lateral and directional effects which must be considered:

a. Spiral instability
b. Oscillatory instability.

Spiral Instability

Suppose an aeroplane is disturbed in bank only. The resulting sideslip produces not only a restoring moment in the rolling plane, but also a yawing effect into the bank because of the directional stability. If the directional stability produces greater yawing effects than the lateral stability produces restoring effects, the result may be an increase in bank rather than a decrease, and the aircraft may enter a diving turn of increasing steepness. Most aircraft have better directional stability than they have lateral stability and are therefore spirally unstable. The effect is normally unimportant except at low airspeeds with asymmetric power, when excessive yaw can very quickly result in an uncontrollable spiral effect.

Oscillatory Instability

The interaction of rolling and yawing effects may produce oscillatory instability. Oscillatory instability is a combination of roll and yaw; if the roll is dominant it is called a *Dutch Roll,*, if the yaw is dominant it is called *Snaking*, and is usually associated with high wing loading, high altitude, sweepback and low EAS. The aerodynamic causes of oscillatory instability are complicated, but a simple explanation of one form of Dutch Roll is given in the following paragraph.

Dutch Roll

Consider a swept wing aircraft which yaws for some reason. The forward starboard wing generates more lift, thus rolling the aircraft to port; the forward wing also, however, generates more drag, because of increase in induced drag, and it will because it presents a larger effective area to the airflow, as seen in Figure 61, so the aircraft yaws to starboard and starts a new cycle of events. The result is a most uncomfortable and possibly dangerous wallowing motion. In many high performance aeroplanes yaw dampers operating usually through the autopilot, are installed to prevent or limit oscillatory instability.

6.
The Aeroplane in Flight

Straight and Level Flight – The Climb – The Dive – The Glide – Take-Off and Landing Approach Case – Turning Flight – High Speed Flight – Autorotation – The Spin

6 THE AEROPLANE IN FLIGHT

Introduction
We have so far considered the basic mechanics and physics of aeronautics, and the manner in which an aeroplane generates various forces which act on it in flight. We have also seen the effects of various artificial devices on the aerodynamics of an aircraft. The time has now come to view the aircraft as a whole in straight and level flight, in the climb and the dive, and also in the turn. We must also consider the control of the aircraft and we will later look at the stability of the aircraft.

Straight and Level Flight

The Aerodynamic Balance
In straight and level flight the forces acting on an aeroplane may be represented as shown in Figure 63 below:

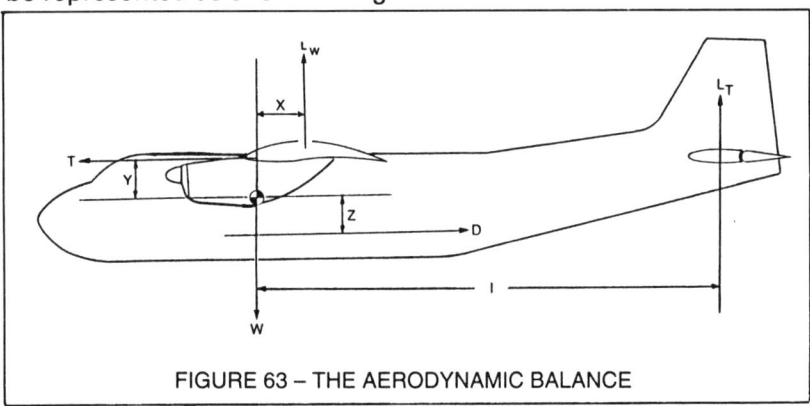

FIGURE 63 – THE AERODYNAMIC BALANCE

These forces must be in equilibrium when resolved in any direction, and the total moment about the C of G must be zero.

For vertical equilibrium $\quad L_W + L_T = W$ .... (1)
For horizontal equilibrium $\quad T = D$ .... (2)
For rotational equilibrium
$\quad\quad L_W \times X + D \times Z = T \times Y + L_T \times l$ .... (3)

The contribution to total lift provided by the tailplane (L_T) is usually very small and is normally neglected in equation (1).
The tailplane moment $L_T \times l$ is however the key to balance in

equation (3); the function of the tailplane is simply to provide the necessary force to prevent rotation (or to *Produce* rotation when required).

Note that an aerodynamic balance is normally achieved automatically in the other two (rolling and yawing) planes because the aeroplane is usually symmetrical in those planes.

The Simplified Aerodynamic Balance
For all further study we can assume that the tailplane is negligible and that the tailplane lift is doing its job properly; this means that lift, weight, thrust and drag will be considered to act through the centre of gravity as illustrated in Figure 64.

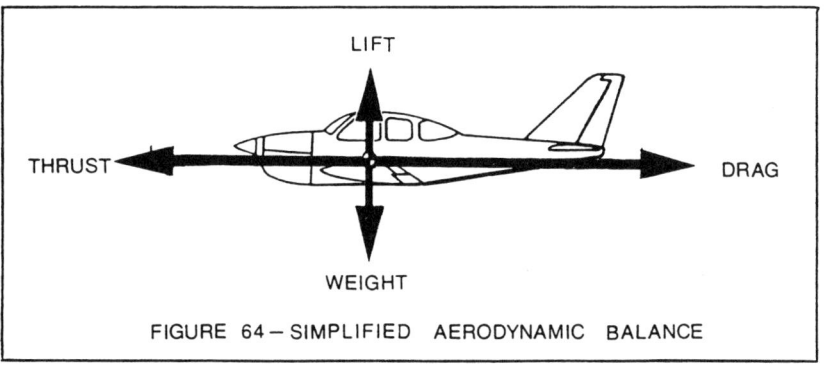

FIGURE 64 — SIMPLIFIED AERODYNAMIC BALANCE

The Climb

Consider an aeroplane climbing at constant speed on a flight path inclined at an angle of ø° to the horizontal.

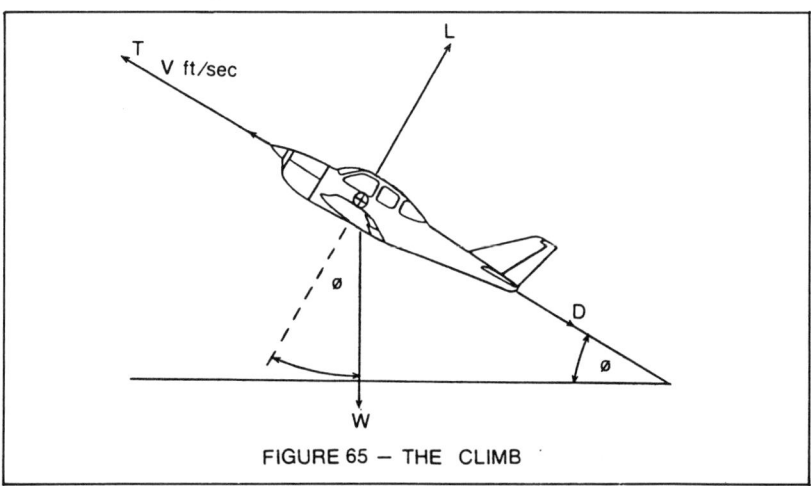

FIGURE 65 — THE CLIMB

The forces acting on the aeroplane will be thrust, drag, lift and weight, disposed as shown in the diagram, Figure 65. We have assumed a simplified aerodynamic balance as we had in Figure 64.

Since the aircraft is in equilibrium on its flight path, forces along and across the flight path must be in balance.

a. Resolving forces along the flight path:
$$T = D + W \sin ø \quad \text{......... (4)}$$
b. Resolving forces across the flight path:
$$L = W \cos ø \quad \text{............... (5)}$$

Equations (4) and (5) simply tell us that in the climb:

a. Lift is less than weight (in a vertical climb lift is zero)
b. Thrust must be greater than drag because the weight component has to be countered.

During a climb an aircraft gains potential energy by virtue of its elevation (or height); this is achieved by one or, a combination of two ways:

a. The usage of propulsive energy above that required to maintain level flight.
b. The usage of aircraft kinetic energy, i.e. loss of velocity by a zoom.

Zooming for altitude is a transient process of exchanging kinetic energy for potential energy and is of considerable importance for aircraft which can operate at very high levels of kinetic energy. However, the main portion of climb performance for most aircraft is a near steady process in which additional propulsive energy is converted into potential energy.

Forces in the Climb

We have briefly mentioned the forces used in the climb and now we can look more deeply into them. To maintain a climb at a given EAS more power has to be provided than in level flight; this is to:

a. overcome the drag as in level flight ($T_R = D \times V$) (i.e. Thrust $_{Req}$ = Drag × Velocity)
b. lift the weight at a vertical air speed, which is known as the rate of climb
c. accelerate the aircraft slowly as the TAS steadily increases with increasing altitude, therefore:
$T_R = DV + WV_c + WV \, a/g$ where V_c is the rate of climb and 'a' is the acceleration.

The acceleration term can be ignored in low performance

aircraft but has to be taken into account in jet aircraft with high rates of climb.

We can see in Figure 66 that the rate of climb is determined by the excess power and the angle of climb is determined by the amount of excess thrust left after opposing drag. It will be seen from this that in the climb the lift force is less than the weight, that is to maintain balanced flight. This is because the excess thrust force supports the weight that is not supported by the lift (W sin ø).

The aircraft could climb straight up (ø = 90°), lift would be zero and the excess thrust would support the entire weight after drag has been overcome, the rate of climb depending upon the excess amount of thrust.

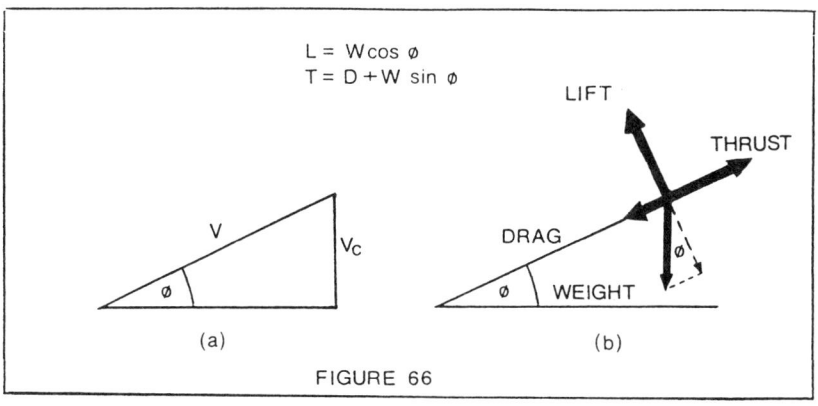

FIGURE 66

In Figure 66 (a): $\sin \emptyset = \dfrac{*\text{rate of climb}}{V}$

In Figure 66 (b): $\sin \emptyset = \dfrac{\text{thrust} - \text{drag}}{\text{weight}}$

Since ø is the same in each case:

$$\frac{\text{rate of climb}}{V} = \frac{\text{thrust} - \text{drag}}{\text{weight}}$$

Therefore, rate of climb $= \dfrac{V(\text{thrust} - \text{drag})}{\text{weight}}$

$= \dfrac{T_{AV} - T_{Req}}{\text{weight}}$

$= \dfrac{\text{excess thrust (power)}}{\text{weight}}$

In practice aircraft do not, for varying reasons, always use the exact speed for maximum rate of climb. In jet aircraft this speed is quite high and at low altitude is not very critical due to the shape of the power available curve. In piston aircraft the speed is much lower and is normally found to be in the vicinity of the minimum drag speed.

When the maximum angle of climb is required it can be seen from Figure 66 (b), where sin ∅:

$$= \frac{\text{thrust} - \text{drag}}{\text{weight}}$$

that the aircraft should be flown at the speed which gives the maximum difference between thrust and drag. For piston aircraft, where thrust is reducing as speed is increased beyond unstick, the best speed is usually as low as is safe above unstick speed. For a jet aircraft, since thrust varies little with speed, the best speed is at minimum drag speed as shown in Figure 67.

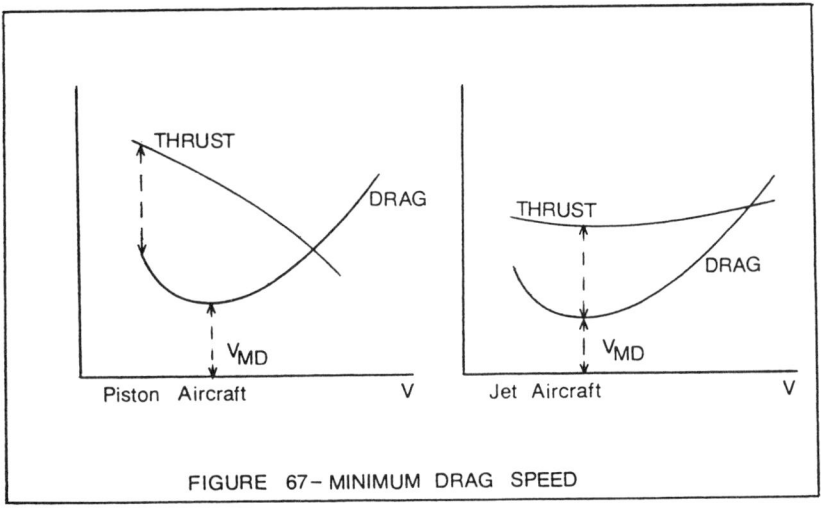

FIGURE 67 – MINIMUM DRAG SPEED

Climbing Performance

The vertical distance between the power available and the power required curves represents the power available for climbing at the particular speed. The best climbing speed (highest rate of climb) is that at which the excess power is at a maximum; so that, after expending some power in overcoming the drag, the maximum amount of power remains available for climbing the aircraft.

For the piston engine aircraft the best speed is seen to be about 180 knots, and for a jet about 400 knots. Notice that in the latter case a fairly wide band of speeds would still give the same amount of excess power for the climb but in practice the highest is used since better engine efficiency is obtained. At the intersection of the curves (points X and Y) all the available power is being used to overcome drag and none is available for climbing; these points therefore represent the minimum and maximum speeds possible for the particular power setting as shown in Figure 68.

If power is reduced the power available curve is lowered. Consequently the maximum rate of climb and maximum speed are reduced, while the minimum speed is increased. When the power is reduced to the point when the power available curve is tangential (i.e. when the straight line touches the curve) to the power required curve, the points X and Y coincide and the aircraft cannot climb, (i.e. $T_{AV} = T_{Req}$).

Ceilings

The altitude at which the maximum power available curve only just touches the power required curve and a sustained rate of climb is no longer possible is known as the *Absolute Ceiling*. It is

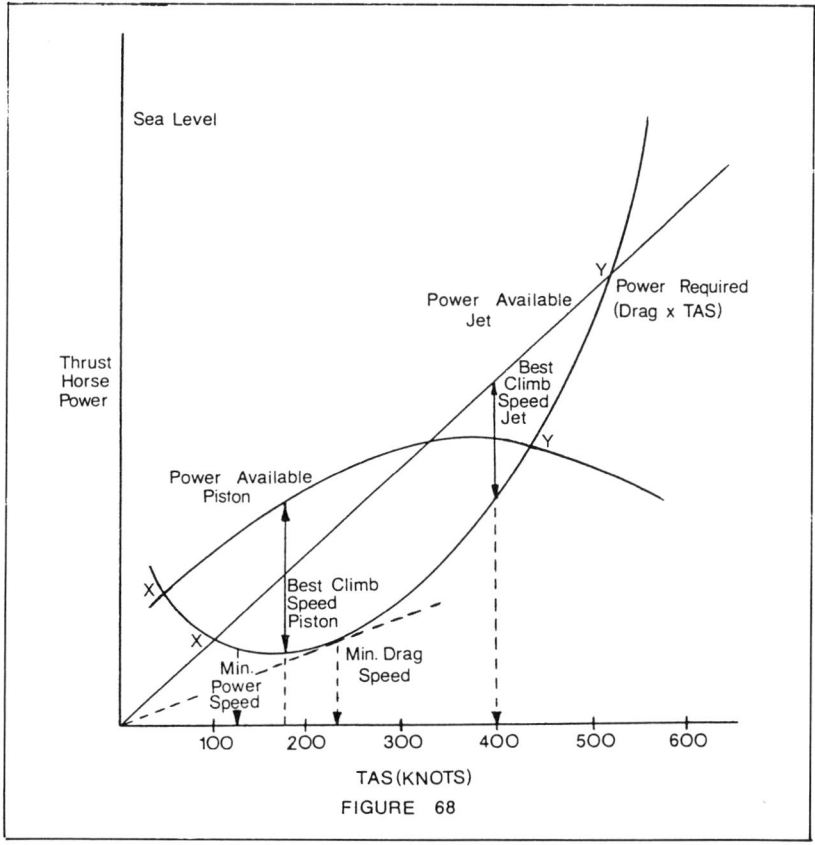

FIGURE 68

possible to exceed this altitude by the zoom climb technique which converts the aircraft's kinetic energy (speed) to potential energy (altitude). Another ceiling is the *Service Ceiling* which is defined as the altitude at which the maximum sustained rate of climb fall to 500 fpm (100 fpm for piston aircraft).

The Dive

Consider an aeroplane diving at a constant speed on a flight path inclined at ø° to the horizontal as shown in Figure 69.

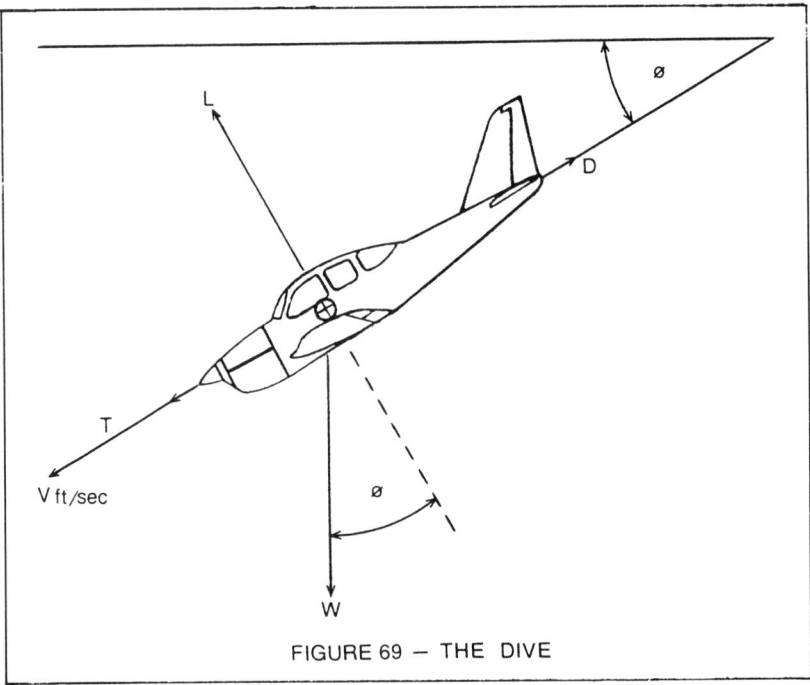

FIGURE 69 — THE DIVE

The relevant forces are again shown and resolution along and across flight path gives:

a. $T = D - W \sin ø$ .... (6)
b. $L = W \cos ø$ .... (7)

Equations 6 and 7 tell us that in the dive, lift is less than weight, and thrust is less than drag (because a component of weight is assisting the thrust).

The Glide

Having discussed the Dive we must now consider the other case of descending flight and this is known as the Glide. There are two conditions that glides are generally considered with relation to aeroplanes and these are:

a. *The Power on Glide* – this condition is used by every aircraft during every flight and is considered normal flight condition
b. *The Power off Glide* – this condition, when thinking of powered aircraft, is normally an EMERGENCY situation due to an engine failure or partial engine failure.

As the first condition is considered a normal condition it is considered much later in aircraft flight performance. The second condition being related to an emergency situation is more important, and we are therefore going to consider this situation at this time.

If we consider the emergency case where we have lost thrust, either due to an engine failure or the fuel supply is exhausted, we can then consider a *Power off Steady Glide* in which case T becomes zero with reference to our equation (6).

This therefore creates the equations as follows:

where $\quad D = W \sin \phi \quad \ldots\ldots\ldots\ldots\ldots\ldots\ldots\ldots\ldots\ldots$ (8)
and $\quad L = W \cos \phi \quad \ldots\ldots\ldots\ldots\ldots\ldots\ldots\ldots\ldots\ldots$ (9)

It now makes our situation more important since the landing approach is made with the throttles closed and therefore has important applications at this stage.

If we divide equation (8) by (9) it gives us

$$\tan \phi = \frac{1}{L/D}$$

The important feature of this last equation is that the *Glide Angle* depends *only* upon the lift/drag ratio, and varies *inversely* with it.

In a power off glide the distance the aircraft will travel over the ground from a given height to ground level, depends upon the *Glide Angle* as illustrated in Figure 70. The flatter this angle, the further the aeroplane will glide over the ground, meaning longer distance. With a little experiment with force vectors, it will soon be seen that the angle is least when drag is at a minimum. When drag is at a minimum the relationship of lift to drag is maximum. To obtain maximum distance, the pilot must fly the aeroplane at the angle of incidence which gives the best lift/drag ratio. The indicated airspeed corresponding to this angle of incidence is worked out by the aeroplane manufacturers.

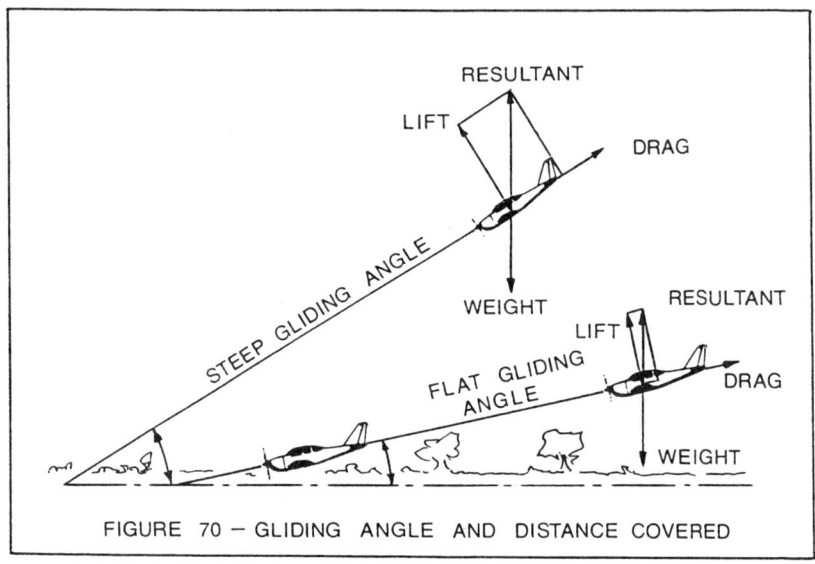

FIGURE 70 – GLIDING ANGLE AND DISTANCE COVERED

In considering the best lift/drag ratio we also obtain the best *Glide Ratio,* and this is defined as:

> The distance an aeroplane can travel over the ground compared to a given height.

This means that the best *Glide Ratio* is the farthest distance an aircraft can glide from a given height, (i.e. if an aeroplane glides 19 miles from a starting height of 1 mile then the glide ratio is 19:1).

Gliding Endurance
For best endurance (time in the air) the aircraft should glide as slowly as possible down the *Optimum Glide Path.* The only way in which speed may be reduced whilst still maintaining the best L/D ratio, is by reducing weight, so that for *Maximum Endurance* the aircraft should have the *Lowest Possible All Up Weight.*

Gliding Range
For best range (longest distance to travel) we need the best glide ratio possible.

Effect of Wind
Since, in a glide for minimum rate of descent the position of the end of the glide is not important, wind will not affect *Endurance.* However, when gliding for range the all-important target is the point of arrival. A *Headwind* will greatly reduce the range of an aeroplane, whereas a *Tailwind* will increase the range.

Example: An aircraft flying at 100 knots that is heading into wind of 100 knots will remain stationary over one position although the angle of incidence remains the same, but would slowly descend.

Effect of Weight
Variation in the weight does not affect the gliding angle provided that the speed is adjusted to fit the AUW. Although the range is not affected by changes in weight, the endurance decreases with increase of weight and vice versa. If two aircraft having the same L/D ratio but with different weights start a glide from the same height, then the heavier aircraft gliding at a higher EAS will cover the distance between the starting point and the touchdown in a shorter time; both will, however, cover the same distance in still air. Therefore the endurance of the heavier aircraft is less.

The Take-Off Considerations

The take-off situation is governed by several conditions such as:
   a. Length of runway available
   b. Obstacles in the line of take-off flight path
   c. Weight of the aeroplane
   d. The power available
   e. Wind direction and speed.

The take-off requirement is mainly to produce enough lift as quickly as possible. The procedure is therefore to apply maximum power (for best acceleration) while increasing the L/D ratio by applying a small amount of flap. The best method of doing this is by applying brakes and opening the throttle to maximum take-off power, and then releasing the brakes. Although this is the ideal way of obtaining maximum power for take-off, it may not be possible due to air traffic control considerations.

The Approach and Landing Case

The landing run after touchdown should be as short as possible, and therefore the approach has to be carefully considered to achieve this, and we find that two distinct requirements are needed for this:

   a. The approach path should, ideally, be as steep as possible to give good obstacle clearance. This ideal is incompatible with modern instrument approach procedures, but does give a shorter landing run, and is therefore adopted and used in Military Tactical situations

b. The speed on the approach should be as low as possible to minimize the landing run.

The approach requirements lead directly to the use of the slats and flaps to raise $C_{L_{max}}$ (thereby reducing V_{min}) and to raise C_D if required (thereby reducing the L/D ratio and steepening the glide).

Slats

For a given angle of incidence, C_L and C_D are fixed so that at the incidence for $C_{L_{max}}$, L/D is fixed and so also is the angle of glide which is very little different from that without the slats.

In addition, the exaggerated nose up attitude of the aircraft spoils the pilot's vision and landing in this attitude often results in the tail touching first and considerable discomfort for the passengers.

Flaps

Flap deflections produce higher values of $C_{L_{max}}$, and also considerably change the L/D ratios. Large flap deflections in particular produce high $C_{L_{max}}$ at low angles of incidence, and considerable reductions in L/D ratio. A large flap deflection thus satisfies *both* requirements of low gliding speed, and steep gliding angle. In addition the angles of incidence involved are so small that the landing attitude is little different from that of the aircraft in level flight. For these reasons flaps are generally used for take-off and landing, (*small* flap deflection for take-off, giving high L/D ratio, *large* flap deflection for landing, giving low L/D ratio), whilst slats are generally used in order to prevent inadvertent stalling.

Turning Flight

Consider an aeroplane performing a steady turn in the horizontal plane. To maintain any body in a turn a centripetal force is needed, i.e. one acting towards the centre of the circle. If this centripetal force is not present the body moves off along a tangent. In an aeroplane the centripetal force is provided by inclining the lift vector by banking the aeroplane, as seen in Figure 71.

If the aeroplane is to describe a steady turn, the centripetal force must be balanced by an equal and opposite force which we call centrifugal force. The force can be shown to be equal to $\frac{WV^2}{gR}$ lbf.

where W = weight of the aeroplane (lbf).
V = horizontal speed of the aeroplane (ft/sec)
g = gravitational acceleration (ft/sec^2)
R = radius of turn (ft).

Thus, in a steady, level turn, with 0° angle of bank, resulting forces horizontally and vertically are:

$$L \sin \phi = \frac{WV^2}{gR} \quad \text{............................} \quad (a)$$

$$L \cos \phi = W \quad \text{............................} \quad (b)$$

and dividing (a) by (b) gives:

$$\tan \phi = \frac{V^2}{gR} \quad \text{............................} \quad (c)$$

So, for a given speed, only one angle of bank will give a given radius of turn. If ø is too large, the aeroplane will sideslip into the turn, effectively reducing the radius. If ø is too small the aeroplane will skid outwards, increasing the radius.

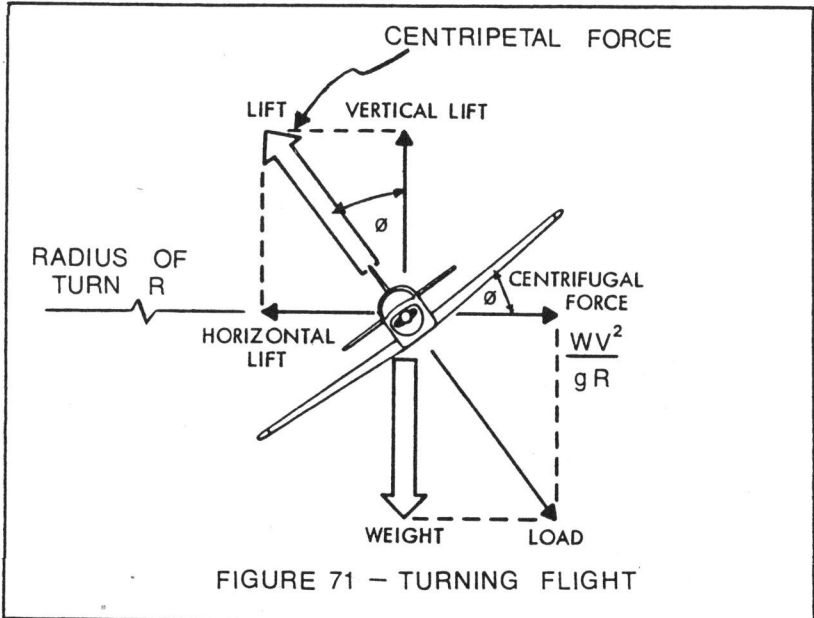

FIGURE 71 — TURNING FLIGHT

Load Factor
In a turn, the lift must increase to balance the *apparent* increase in weight caused by the centrifugal effects. Consider the diagram in Figure 72.

Aerodynamics for the Professional Pilot

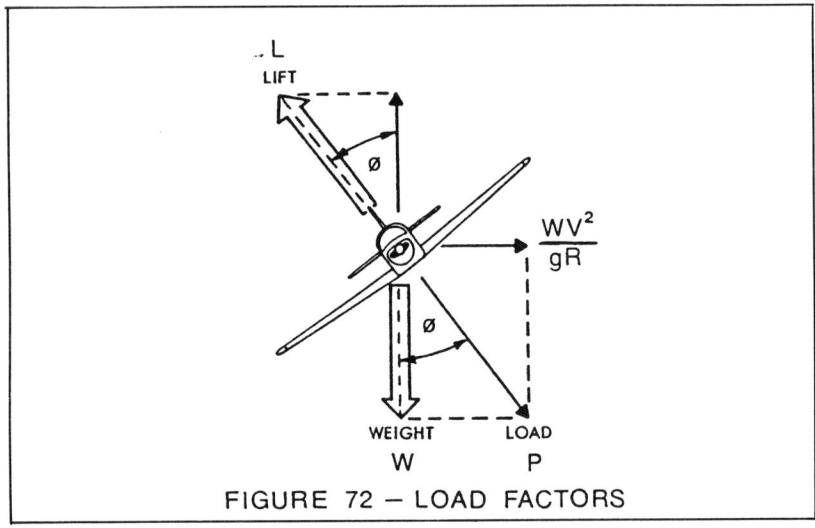

FIGURE 72 — LOAD FACTORS

The force which lift has to counter is the sum of the components of weight and centrifugal force acting normal to the flight path.

From which: $$L = \frac{W}{\cos \phi}$$

Which tells us that in the turn lift is greater than the real weight by a factor of
$$\frac{1}{\cos \phi}$$
This means that the loads on the aircraft are increased by this factor, known as the Load Factor.

So, $$\text{Load Factor} = \frac{L}{W} = \frac{1}{\cos \phi} \text{ (for } bank \text{ only)}$$

The load factor is popularly referred to as the 'g' pulled in the turn, and leads to the perhaps surprisingly point that the amount of 'g' in a turn depends *only* on the angle of bank.

Minimum Flying Speed in a Turn
In straight and level flight

$$V_{minL} = \sqrt{\frac{2 \times L}{C_{L\,max}\, \rho S}} = \sqrt{\frac{2 \times W}{C_{L\,max}\, \rho S}}$$

In a turning flight

$$V_{minT} = \sqrt{\frac{2 \times L}{C_{L\,max}\, \rho S}} = \sqrt{\frac{2 \times W}{C_{L\,max}\, \rho S \cos \phi}}$$

$$V_{minT} = V_{minL} \times \frac{1}{\cos \emptyset}$$

Thus is 2g is pulled (60° of bank), the minimum flying speed increases by a factor of 1.414.

Looping Manoeuvre
We have so far considered a turn in the horizontal plane only. A turn in the vertical plane, i.e. a loop or pull-out from a dive, gives us a similar result.

Consider a pull-out from a dive as shown in Figure 73. Assuming constant speed and a circular arc, both most unlikely, we have

$$L = W \cos \emptyset + \frac{W V^2}{gR}$$

but note that this is a changing condition because the angle $\emptyset$ is changing. This shows that lift must again increase to counter the centrifugal effects. The load factor is increased; 'g' is pulled as shown in Figure 73.

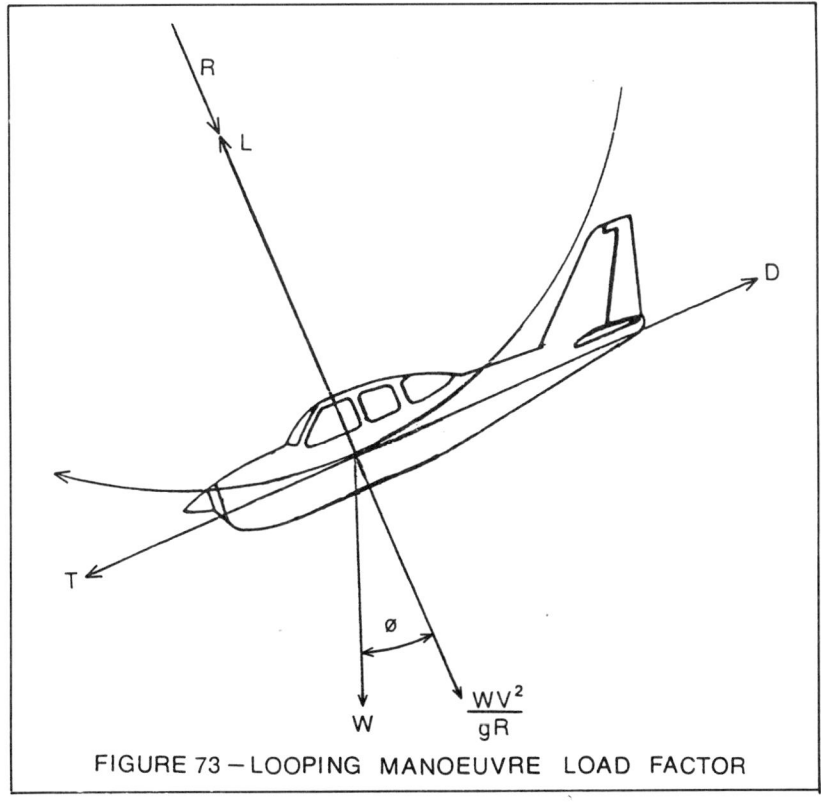

FIGURE 73 – LOOPING MANOEUVRE LOAD FACTOR

High Speed Flight

High Speed Flight
The major difference between the aerodynamics of *High Speed Flight* and the conventional low speed aerodynamics with which we have so far been concerned, is that we now have to take into account compressibility effects. When this is done, it is found that for Dynamic Similarity, a new nondimensional number becomes important and this is known as the *Mach Number*.

Compressibility Effects in Aerodynamics
A compressible fluid is defined as one in which *changes of pressure* are accompanied by *changes in density*.

By this definition all gases, and in particular air, are compressible. However, at moderate speeds (up to 300–400 mph) air may be considered as incompressible, because the density changes which take place around an aircraft are relatively small, and can be ignored. This is because the disturbance created by the aircraft are propagated as pressure waves which travel through the air at speed which is great compared with the speed of the aircraft. When the speed of the aircraft becomes comparable with the speed of propagation of the pressure waves, the compressibility of the air can no longer be ignored, as the pressure disturbances build up around the aircraft and produce appreciable density changes.

The Speed of Sound in a Fluid
A pressure disturbance, or pressure wave, travels through a fluid as a series of compressions and rarefactions, and the mean position of each fluid particle does not change with the passage of disturbance. This is called a 'longitudinal wave form' and the obvious example is a sound wave. (It may be compared with a 'lateral wave form', e.g. a ripple on a pond etc.). In air, the individual molecules are in constant motion in random directions with a mean speed of about 1700 ft/sec (1000 knots) at sea level. Thus the speed of a pressure disturbance in a fixed direction will be somewhat less, about 1100 ft/sec (661 knots) at sea level.

It has been found, in experiments, that *all* pressure disturbances in which the pressure changes are *small*, travel at the same speed, and although the disturbances may not be audible, this speed is the *speed of sound* in the fluid.

All pressure disturbances are affected by the following:

Temperature
Density
Specific Heat
Mass.

Since at different heights the absolute temperature is different, then the speed of sound will also vary with height. According to the International Standard Atmosphere this variation being shown in the following table:

HEIGHT (ft)	T (°K)	SPEED OF SOUND		
		ft/sec	mph	knots
0	288.0	1117	761.6	661
5 000	278.1	1098	748.4	650
10 000	268.2	1078	734.9	638
15 000	258.3	1058	721.2	626
20 000	248.4	1037	707.3	613
25 000	238.5	1017	693.1	602
30 000	228.6	995	678.5	588
35 000	218.7	973	663.7	576
36 090 and above	216.5	968	660.3	573

Definitions

Before proceeding further it is necessary to clearly define certain terms.

Speed of Sound — The speed at which a very small pressure disturbance is produced in fluid under specified conditions.

Mach Number — The ratio of the speed of an object or flow to the speed of sound in the same part of the atmosphere.

Free Stream Mach Number (M_{FS}) — This is the Mach number of the flow sufficiently remote from the aircraft to be unaffected by it.

i.e. $$\text{Mach Number} = \frac{\text{TAS}}{\text{local speed of sound}}$$

M_{FS} is sometimes called the flight Mach number. Ignoring small instrument errors, M_{FS} is the true Mach number of an aircraft as shown on the machmeter.

Local Mach Number (M_L) — When an aircraft flies at a certain M_{FS}, the flow is accelerated in some places and slowed down in others. The speed of sound also changes because the temperature around the aircraft changes. Hence,

$$M_L = \frac{\text{speed of flow at point}}{\text{speed of sound at the same point}}$$

M_L may be higher than, the same as, or lower than M_{FS}.

Critical Mach Number (M_{crit}) — As M_{FS} increases, so do some of the local Mach numbers. That M_{FS} at which any M_L has reached unity is called the critical Mach number. M_{crit} for an aircraft or wing varies with the angle of attack; it also marks the lower limit of a speed band wherein M_L may be either subsonic or supersonic. This band is known as the transonic range.

Definitions of Flow

Changes in airflow occur at $M_L = 1.0$ and the boundary between each region of flow is that M_{FS} which produces an M_L appropriate to that region as shown in Figure 74.

Notes:
1. The subsonic region has been subdivided at $M = 0.4$ since below this Mach number errors in dynamic pressure, assuming incompressibility, are small. However, compressibility effects can be present even at $M = 0.4$; whether they are or not depends on wing section and angle of attack.
2. The actual values of M_{crit} depend on individual aircraft and angle of attack.
3. Another definition of the transonic range is: that range M_{FS} during which shockwaves form and move significantly.

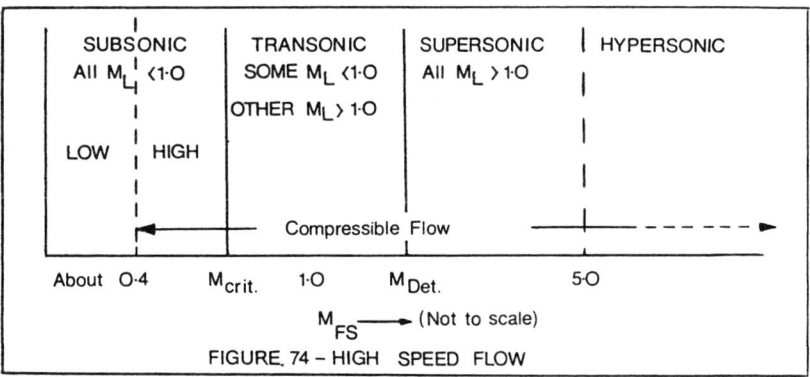

FIGURE 74 – HIGH SPEED FLOW

Flow Through a Stream Tube

In order to investigate the effects of compressibility, or even consider them, it is necessary to consider the basic laws which govern fluid flow as illustrated in Figure 75.

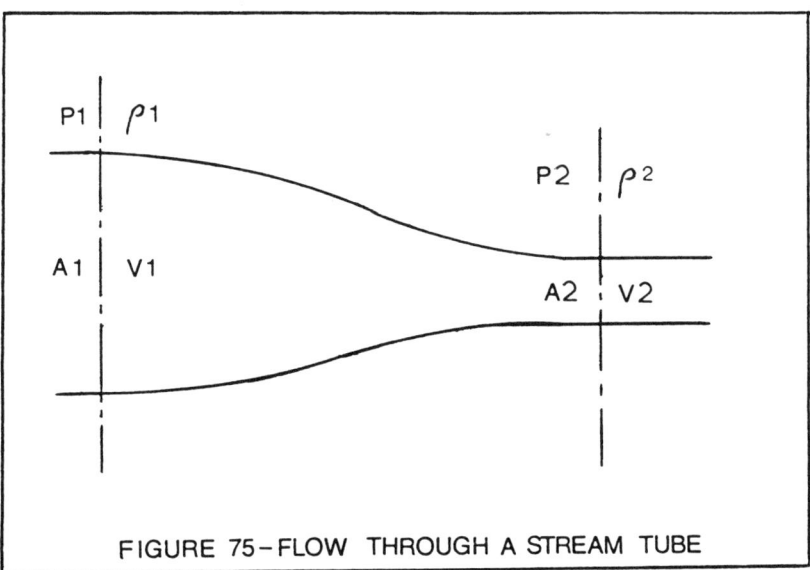

FIGURE 75 – FLOW THROUGH A STREAM TUBE

These laws are:
1. The law of continuity of mass flow

   i.e. $Av = $ constant or $\dfrac{d\rho}{\rho} + \dfrac{dv}{v} + \dfrac{dA}{A} = 0$

2. Bernoulli's Equation

   i.e. $vdv + \dfrac{dp}{\rho} = 0$

3. The Adiabatic Gas Law

   i.e. $pV = $ constant or $\dfrac{p}{\rho^\gamma} = $ constant C

4. The expression for the speed of sound

   $a = \sqrt{\dfrac{\gamma p}{\rho}}$ or $a^2 = \dfrac{\gamma p}{\rho}$

Note: γ is the ratio of specific heats $\dfrac{Cp}{Cv}$

The mathematical equations results can be summarized as follows:

	SUBSONIC FLOW	SUPERSONIC FLOW
CONTRACTING CHANNEL	Accelerates, and static pressure decreases	Decelerates, and static pressure increases
EXPANDING CHANNEL	Decelerates, and static pressure increases	Accelerates, and static pressure decreases

Thus for acceleration up to sonic speed a *convergent* duct is required whilst for acceleration *above* sonic speed, a *divergent* duct is required.

The Propagation of Pressure Disturbance from a Point Source
(A point source is a body whose size tends to be zero)

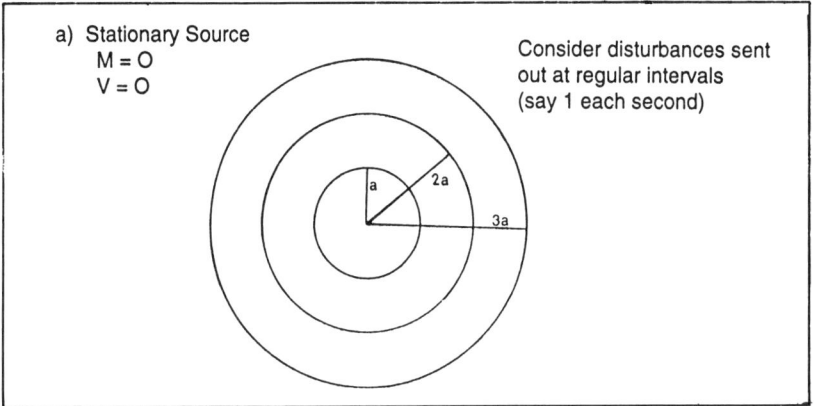

a) Stationary Source
  M = O
  V = O

Consider disturbances sent out at regular intervals (say 1 each second)

Pressure waves will be sent out continuously – the radius of the sphere (circle in two dimensions) increasing at the speed of sound.

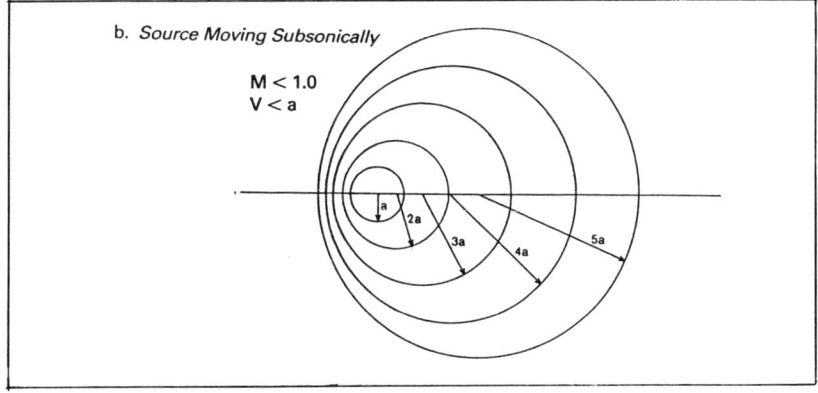

b. *Source Moving Subsonically*
  M < 1.0
  V < a

When moving at subsonic speed, the air ahead of the point source still receives advance warning of the approach of the point source. (Notice that the frequency depends on the position of the observer to the direction of motion. This example explains the high pitch of the sound of a train whistle as it approaches and the relative low pitch as the train travels away into the distance.)

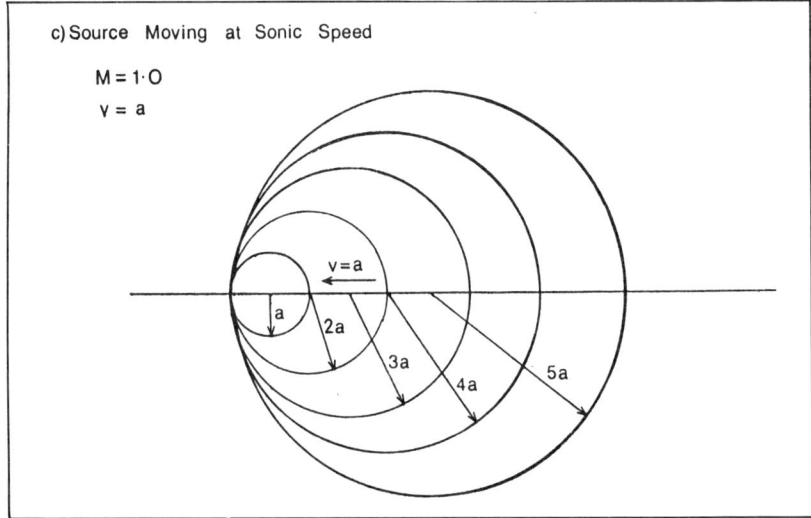

In the sonic case the waves move along with the source leading to an intense compression wave. Therefore in line with the motion, one would hear a crack, as of a whip, but one side would hear a high pitched whistle.

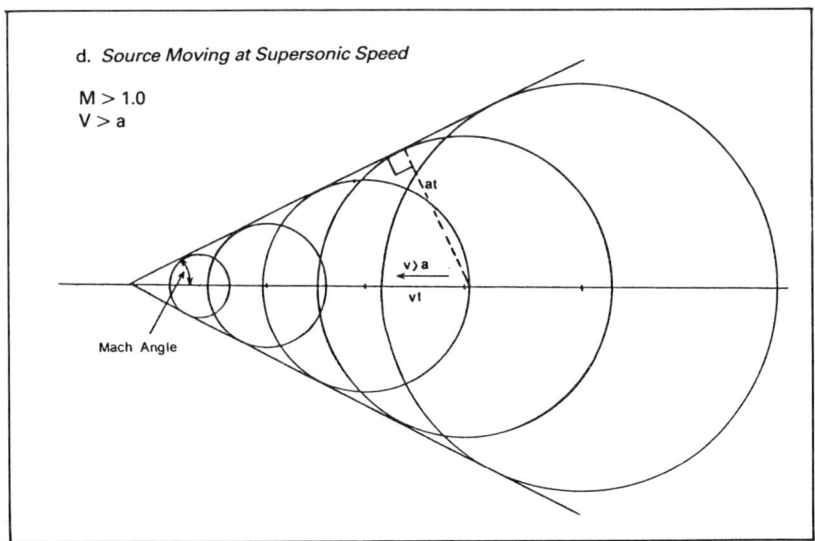

In the supersonic case, the influence of the point source is confined to a cone with itself as the apex and with generators inclined backwards, making an angle with the direction of motion known as the *Mach Angle*. Ahead of the cone the air is unaffected by the presence of the source of disturbance. Thus the waves at any instant form a wave front known as *Mach Wave* or *Mach Cone*. The generators of the cone are *Mach Lines*.

Shock Wave Formation

At M_{crit}, there is *one point* on the aerofoil at which the local Mach Number, $M_L = 1.0$. Pressure waves arising at points on the surface of the aerofoil downstream will be able to make no further progress upstream when they reach this point. They will thus 'pile-up' at this point and the shock wave will begin to form, starting as a 'point' shock wave as illustrated in Figure 76.

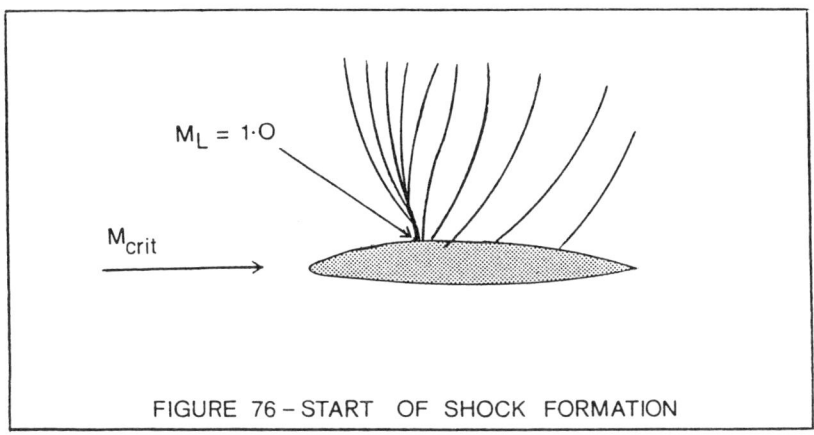

FIGURE 76 – START OF SHOCK FORMATION

With increase of the free stream Mach number above M_{crit} a supersonic region forms on the wing with the shock wave at the downstream boundary. With further increase in the free stream Mach number, the shock wave on the upper surface grows in strength and moves backwards towards the trailing edge. Eventually, a similar shock wave will form on the lower surface in exactly the same manner, and this also will grow in strength and move back.

This process is illustrated in Figure 77 with the aerofoil at a small positive incidence and the *Free Stream Mach* number gradually increasing:

Note: Because of the flow separation caused by the upper shockwave, due to greater pressure difference, the lower shock wave reaches the trailing edge first.

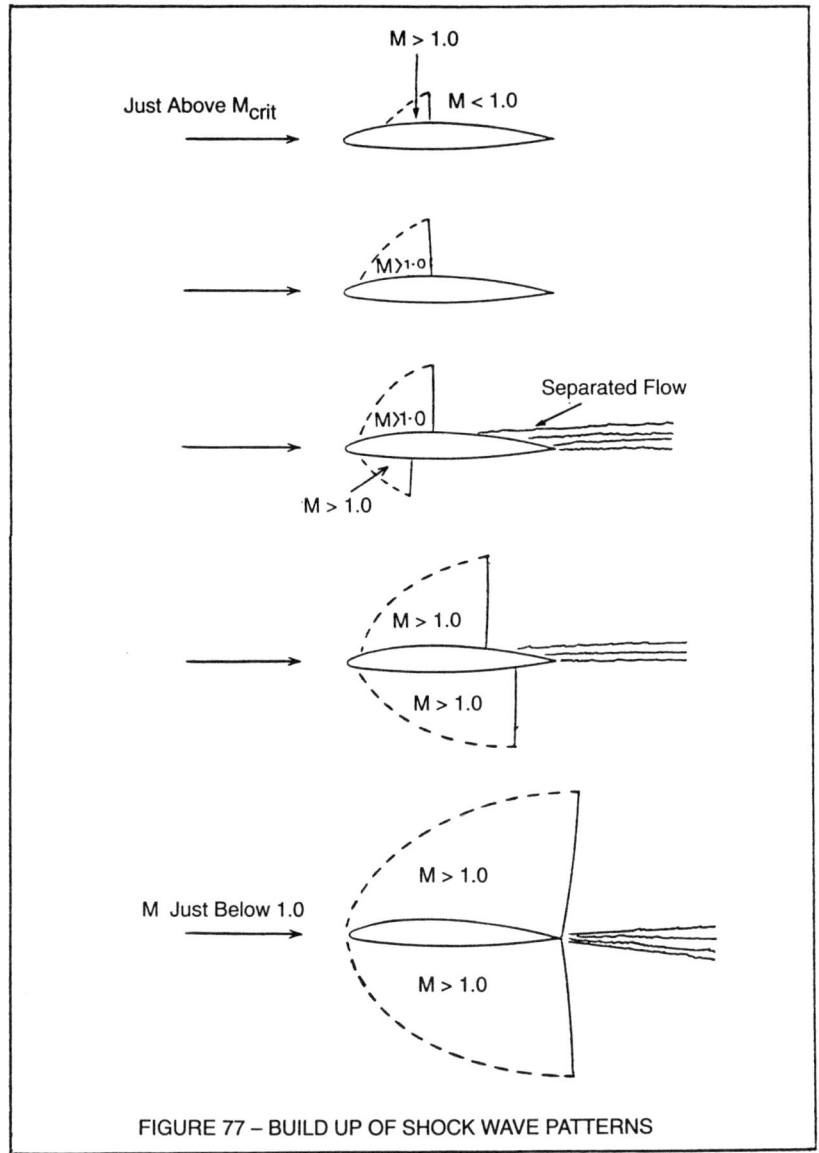

FIGURE 77 – BUILD UP OF SHOCK WAVE PATTERNS

Flow Through Shock Waves

A shock wave is defined as a thin region across which take place sudden changes in the velocity, pressure, density, and temperature of the gas passing through it.

There are two types of shock wave as illustrated in Figure 78:

a. *Normal Shock Wave,* i.e. perpendicular to the stream
b. *Oblique Shock Wave,* i.e. inclined to the stream.

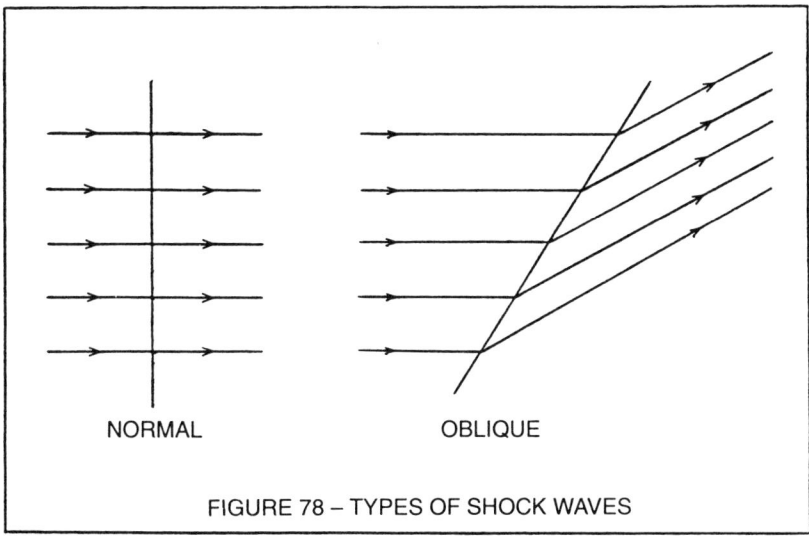

FIGURE 78 – TYPES OF SHOCK WAVES

a. *Normal Shock Wave*
When a streamline travels across a normal shock wave, *there is no change in the direction of the flow,* but the following changes take place:

1. A reduction in speed (always from *supersonic* to *subsonic*)
2. An increase in density
3. An increase in static pressure
4. A rise in temperature.

b. *Oblique Shock Waves*
When a streamline travels across an oblique shock wave, *the flow is deflected,* and in addition, the following changes take place:

1. A reduction in speed (but usually still *supersonic* behind the shock)
2. An increase in density
3. An increase in static pressure
4. A rise in temperature.

Aerodynamic Effects of Shock Waves
When a 'Normal' shock forms on the surface of an aircraft wing, due to the high pressure difference across the shock wave, the boundary layer starts to thicken ahead, and will eventually separate at the attachment point of the shock wave. This is called 'shock induced separation', and it leads to effects similar in many ways to the low-speed high-incidence stall. However, the 'shock stall' as these effects are called, can occur at *any* angle of attack,

provided the speed is right, unlike the low-speed stall, which occurs only at the 'stalling angle' and the corresponding stalling speed.

The 'shock stall' consists of:

1. Rapid increase in C_D
2. Sudden decrease in C_L
3. Large and erratic changes of pitching moment.

General Pattern of Aircraft Behaviour at High Mach Numbers

This may be summarized as follows:
a. In the Transonic Range:
   1. Changes in longitudinal stability, which may take the form of:
      a. nose-up or nose-down changes of trim
      b. 'porpoising' and longitudinal pitching
   2. Changes in lateral stability:
      a. alternate wing dropping
      b. a single wing dropping severely.
   3. Changes in directional stability:
      a. directional oscillations with poor or no damping ('snaking')
      b. a sudden directional jerking or kicking over of the rudder.
   4. Severe buffeting:
      a. of some main part of the structure
      b. of control surface.
b. In the Supersonic Range:
   1. Cessation of buffeting and general smoothing-out of the flight
   2. Return to more normal stability.
   3. A general 'heavying' of controls with a great loss of effectiveness, i.e. very low rates of roll, and lack of elevator power.

Design Methods of Reducing Adverse High Speed Effects

Design methods of reducing high speed effects are:

   a. Sweepback
   b. Reduced thickness-chord ratio
   c. Boundary Layer Control
   d. All-moving tailplanes
   e. Power operated controls
   f. Area rule.

a. *Sweepback*

This was first suggested by German aerodynamicists about 1935. Before this, sweepback had only been used for stability reasons. Nowadays it is used to raise M_{crit} and/or reduce the effects of a shock stall, as well as to reduce drag at supersonic speeds. As in the case of stability, with a swept wing, it is only the component of air velocity *Perpendicular* to the leading edge which determines the forces acting and is modified by the wing profile as illustrated in Figure 79. Therefore if a straight wing has a critical Mach number denoted by $(M_{crit})_{st}$ and the wing is then swept through an angle ø, the new critical Mach number $(M_{crit})_{sw}$ will be given by

$$(M_{crit})_{sw} = \frac{(M_{crit})_{st}}{\cos ø}$$

In theory, sweep*forward* should be just as effective in raising M_{crit} as sweepback. However, sweepforward is not normally used because of longitudinal stability considerations and danger of 'wing divergence'.

Note: Sweepback not only raises M_{crit}, but also reduces the effects of the 'shock stall' when it *does* occur.

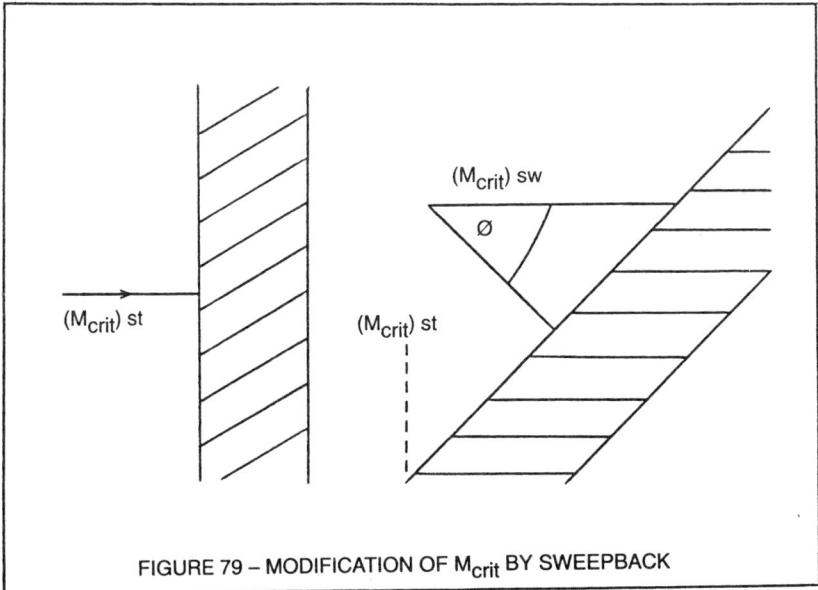

FIGURE 79 – MODIFICATION OF M_{crit} BY SWEEPBACK

b. *Reduced Thickness-Chord Ratio*

The thickness – chord ratio (t/c) is a measure of the aerodynamic thickness of the wing, and is expressed as a percentage. Because a thinner wing causes less speeding up of the air over the upper surface, it will raise the value of M_{crit} and will also reduce the strength of the 'shock stall' when it *does* occur. So considerably thinner wings are now used — for low speed aircraft 15–20% is possible, but 'Hunter' 8%, FD2 4%, F-104 3%, etc.

c. *Boundary Layer Control*
   1. Vortex generators
   2. Blowing
   3. Suction.

The purpose of any form of boundary layer control for *high* speeds is to move the main shock waves back towards the trailing-edge, thus reducing the effect of shock-induced flow separation and the adverse effects connected with it. Vortex generators, as illustrated in Figure 80, achieve this result by setting up vortices which stir fast-moving air from the main stream into the sluggish boundary layer. They are attached in rows to the wing surface, slightly inclined to the flow direction.

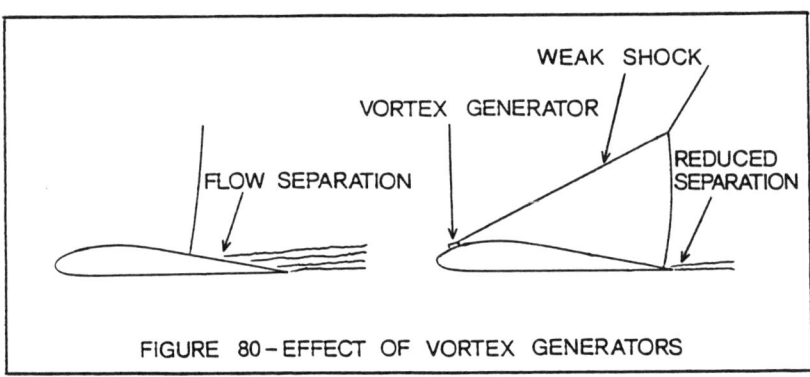

FIGURE 80 – EFFECT OF VORTEX GENERATORS

The other two methods, still very much in the experimental stage, may be used more widely in the future.

d. *'All-moving' or 'All-flying' Tailplanes*
This type of tailplane overcomes the loss of effectiveness of conventional elevators at transonic speeds. The tailplane is made the primary control surface and is coupled directly to the control column. These are usually power operated.

On some aircraft the elevator is retained and is linked to the tailplane in such a way that its movement assists the action of the tailplane. When no elevator is used the tailplane is known as a 'slab' tailplane (F-104, F-101).

Sometimes a variable-incidence (VI) tailplane is used. This is primarily intended as a means of trimming the aircraft, but it does also help to improve control at high Mach numbers. The disadvantage is that movement of the control column must be supplemented by the operation of a separate switch, which is spring loaded to a central off-position. The method of operation of the VI tailplane is usually electric, e.g. 'Canberra'.

e. *Power-Operated Controls*
These are generally necessary at transonic and supersonic speeds. Sometimes the controls are fully power-operated and sometimes power assisted, i.e. the usual stick forces applied by the pilot are augmented mechanically.

Servo-control units may be hydraulic or electrical (normally hydraulic).

Artificial feel is necessary if the controls are fully power-operated, and this is provided either by a spring mechanism or by a bellows arrangement in which the pressure is dynamic pressure ($= \frac{1}{2} \rho V^2$). The movement of the pilot's control compresses the bellows so that the force transmitted to the pilot is proportional to $\frac{1}{2} \rho V^2$. This is called a 'q' feel system since 'q' is the symbol for $\frac{1}{2} \rho V^2$.

f. *Area Rule*
By careful design the total wave drag of the aircraft can be made less than total of the wave drags of its separate components: i.e. fuselage, wings, etc.

In very simple terms, the aircraft is designed so that its cross-sectional area varies along its length approximately like that of the ideally-shaped body of revolution which has minimum wave-drag.

This means that the fuselage has to be 'waisted' at the attachment of the wings, etc. in order to give the required smooth variation of cross-sectional area.

Above the M_{crit} the wave drag caused by the shock waves makes the drag rise rapidly. The wave drag is kept small by making the fuselage long and thin, with a pointed nose and by making the fuselage *Area-Ruled,* as shown in Figure 81.

Aerodynamics for the Professional Pilot

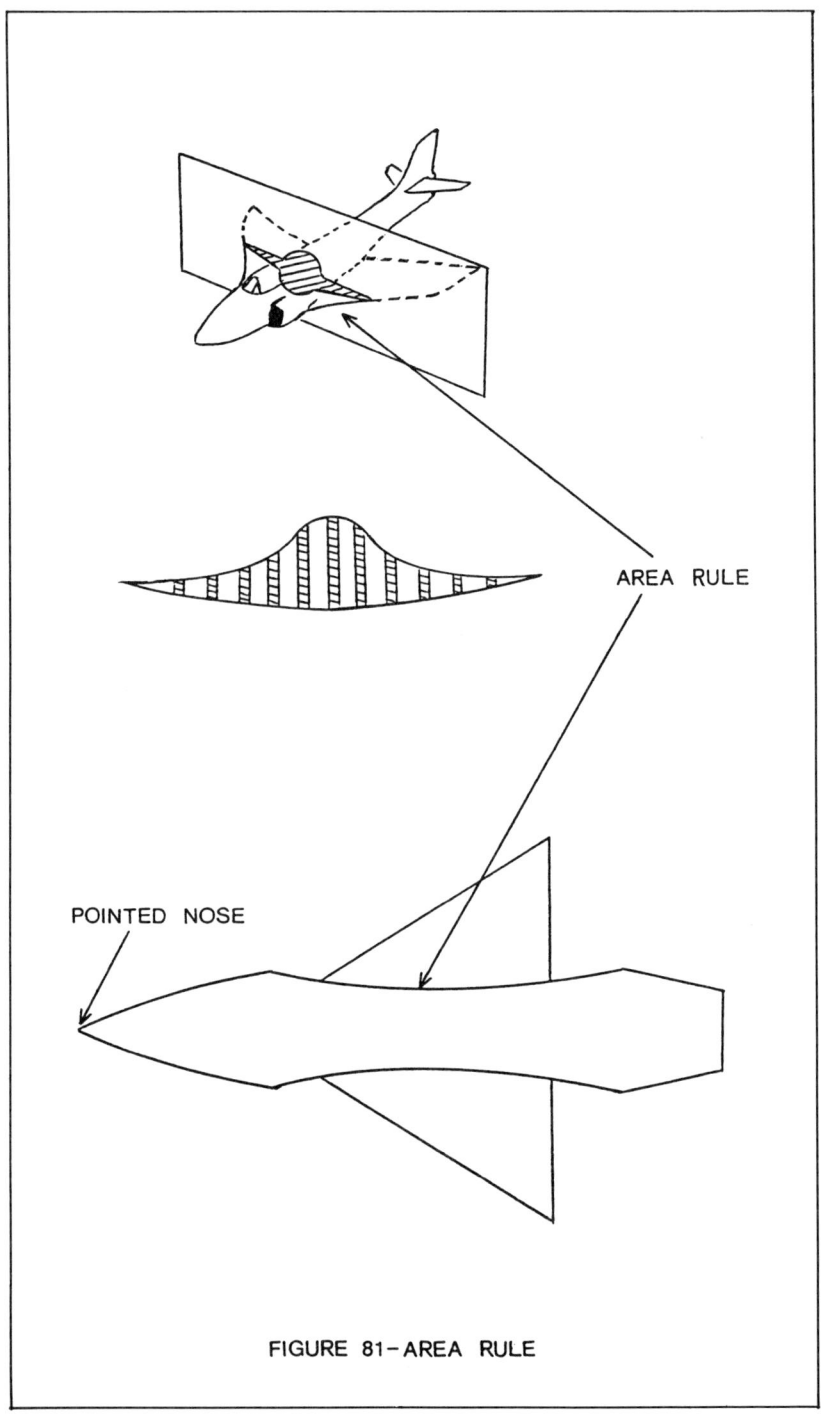

FIGURE 81-AREA RULE

Aerodynamics for the Professional Pilot

Summary

Having now studied the effects of shock waves with increase in Mach numbers, Figure 82 shows the types of shock wave patterns.

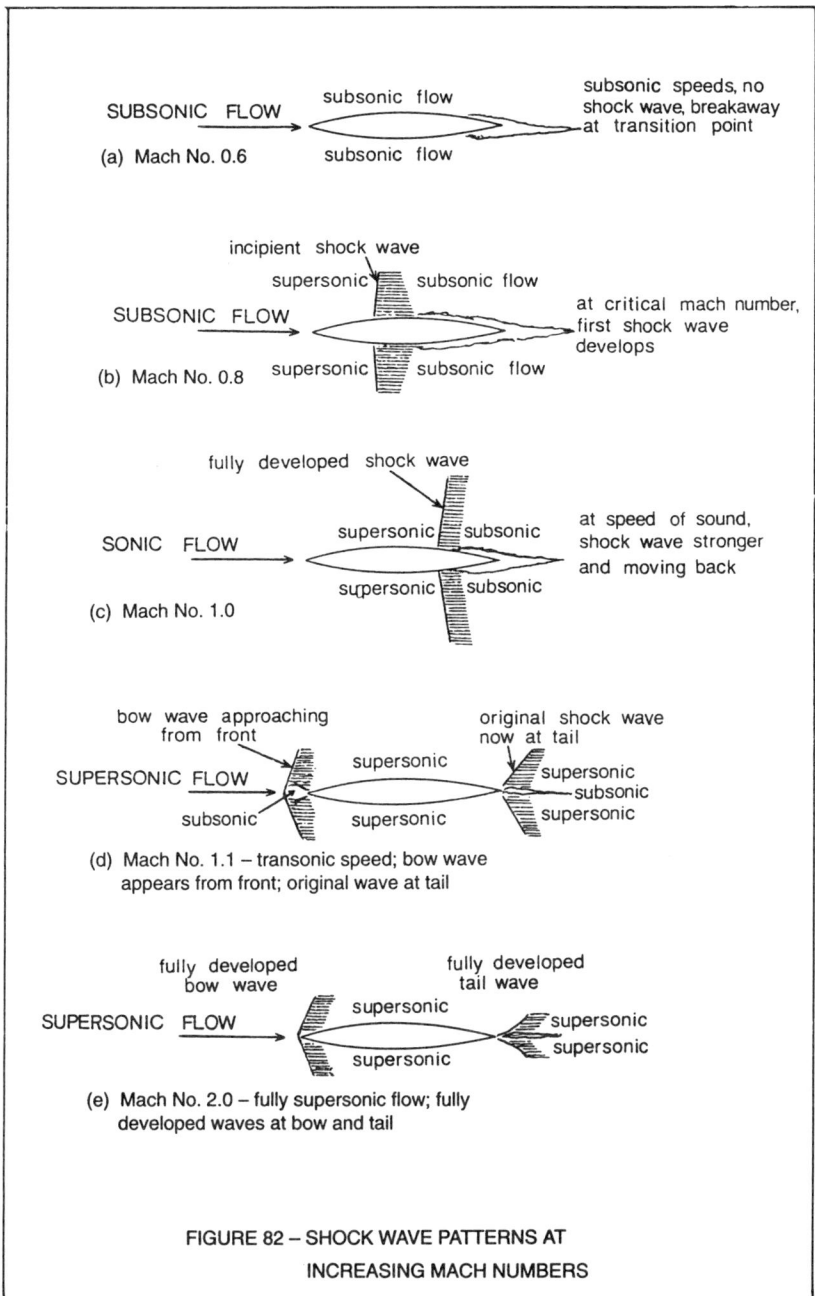

FIGURE 82 – SHOCK WAVE PATTERNS AT INCREASING MACH NUMBERS

Autorotation – The Spin

Introduction
The final manoeuvre which we must consider is the state of autorotation which leads to the spin. Consider an aircraft at high angle of incidence approaching the stall, and if for some reason that aircraft suffers a disturbance in roll, the downgoing wing increases its angle of incidence, whilst that of the upgoing wing is decreased. The downgoing wing stalls, roll continues and the aircraft yaws into the roll because of the sharp increase in drag on the stalled wing. The result is usually a combined rolling, yawing and pitching motion at low EAS about the vertical axis and is known as a spin. The spinning properties of aircraft vary enormously between types and in general only small training and fighter aircraft are cleared for spinning due to the structural stresses involved.

Spinning
The aircraft spin manoeuvre can be dangerous in any aircraft and is normally only performed on small training aircraft or fighter aircraft. It is used on training aircraft for student pilots to see and understand the environment that leads up to the spin and also to learn the spin recovery techniques. Fighter aircraft may go into a spin deliberately for evasive action during aerial combat.

Each type of aircraft has its own mannerisms with reference to spinning under the same type of conditions, so a basic generalization is given in the following paragraphs with a few main points being mentioned for the pilot to look for when converting to different types of aircraft. The inverted spin is another condition to be considered and is discussed later in this chapter. Although spinning should be avoided where possible, all the observations given are based on a deliberately-induced, erect spin to the right.

In the diagrams and following text on spinning there are various signs and phrases which are used. Table 7.1 shows a reference chart to enable the reader to understand these signs and phrases and to allow the reader to mentally visualize the pilot's position and phase of flight.

Phases of the Spin
There are three basic phases of the spin manoeuvre which are:
   a. The Incipient Spin
   b. The Fully Developed Spin
   c. The Recovery.

Autorotation Reference Chart

AXIS (Symbol)	LONGITUDINAL (x)	LATERAL (y)	NORMAL (z)
Directional Position	Forwards	To Right	Downwards
ANGULAR VELOCITY			
Designation	Roll	Pitch	Yaw
Symbol	R	P	Q
Positive Direction	to right	nose up	to right
Moments of Inertia	A	B	C
MOMENTS			
Designation	rolling moment	pitching moment	yawing moment
Symbol	L	M	N
Positive Direction	to right	nose up	to right

Table 1.

The Incipient Spin

This stage lasts for some 2–6 turns which occur in the initial stages. The spin is most likely to occur when the angle of incidence over both wings varies causing one wing to stall. This is most likely to happen:

   a. at low airspeeds, close to the stall, when a crosswind causes strong yawing moments with subsequent loss of lift on one wing
   b. when performing a roll, or rolling movement (i.e. a steep turn right or left) at low airspeeds close to the stall.

The incipient spin is the initial unstable conditions that occur prior to the aircraft entering a steady stable spin.

The Fully Developed Spin

At this stage only the Steady Stabilised Spin will be considered. Although there will be some sideslip, the Steady Stabilised Spin is considered as being when the aircraft has settled down to a steady rate of rotation and a steady rate of descent about the spin axis.

This situation will only develop if the aerodynamic and inertia forces and moments can achieve a state of equilibrium. The attitude of the aircraft at this stage will depend on the aerodynamic shape of the aircraft, the position of the controls and the distribution of mass throughout the aircraft. The centrifugal force acting around the spin axis will also affect the turning moments due to the distribution of mass in the aircraft.

When a spin occurs, there are two components of motion which act on the centre of gravity:

a. The rate of descent of the aircraft which is shown on the VSI is the linear velocity acting in a vertical direction

b. The horizontal motion about the vertical axis, otherwise known as the *Spin Axis,* is normally a small motion. This is determined as the radius of the spin (R), which is the distance between the spin axis and the CG. This distance is small and is approximately half the span of the wing.

When these two motions are combined, the resultant effect is similar to a vertical spiral line the angle of which is usually small (i.e. less than 10°). The motions of the aircraft in a spin is illustrated in Figure 83.

With relation to its CG, an aircraft has turning moments about all three axes when it goes into a spin. It should be noted at this stage that if an aircraft is rotated around *one* of its axes at a high velocity, a gyroscopic force takes place around that particular axis. As there are turning moments about all three aircraft axes, the aircraft is subject to strong gyroscopic action in all three planes of motion. This point should strongly be noted when the recovery action is considered, thereby applying another turning moment to the aircraft's plane of spin.

To understand the forces acting on all three aircraft axes, we must study the relationship between the angular velocities and the aircraft attitude about the *Vertical Spin Axis.*

a. NORMAL AXIS – If the aircraft were to rotate around the spin axis at the same rate as the aircraft rotates around its normal axis it would be seen that the inner wing tip must present itself towards the axis of rotation. This produces a yawing angular motion in the direction of spin (i.e. a turn to the right as illustrated in Figure 83)

Aerodynamics for the Professional Pilot

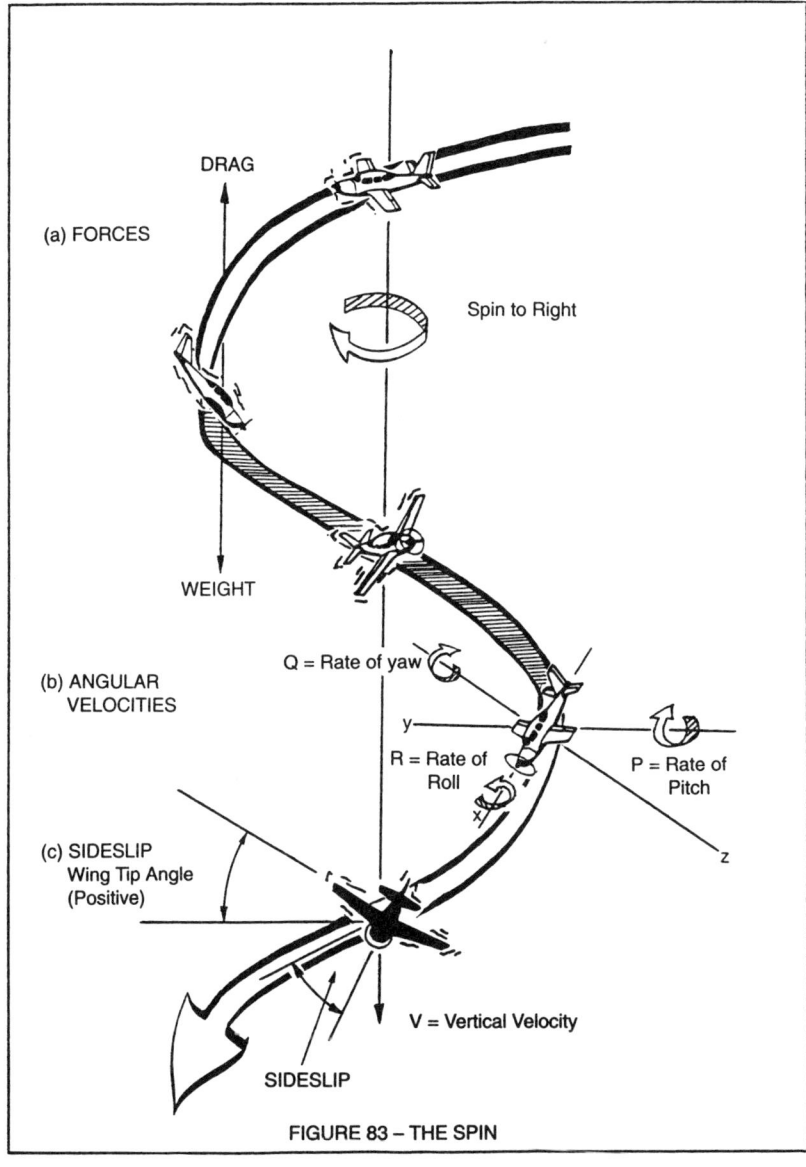

FIGURE 83 – THE SPIN

b. LATERAL AXIS – To enable the same point (the cockpit or the pilot's head) to be presented to the spin axis, the aircraft must rotate about the lateral axis in a nose-up pitching motion. (*Note:* A nose-down pitching motion would increase the radius from the spin axis and create a divergent spinning circle and is dangerous, where recovery could become impossible and the aircraft become inverted.)

c. LONGITUDINAL AXIS – The forces acting on the aircraft to enable it to remain around the spin axis creates an angular motion which will be all roll.

The rate of descent of the aircraft in a spin will be dependent upon the mass forward or aft of the CG. This will determine the angle of pitch of the aircraft and therefore whether a steep spin or a flat spin occurs (Figure 84).

1. Mass forward of the CG – if the larger mass is forward of the CG, then the angle of pitch is greater causing a Steep Spin to occur.
2. Mass aft of the CG – if the larger mass is aft of the CG, then the angle of pitch is smaller and a Flat Spin will occur.

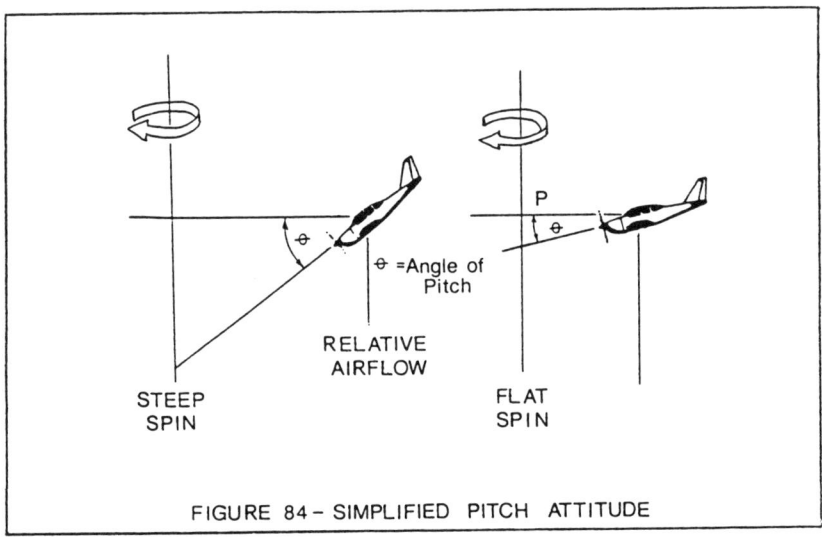

FIGURE 84 – SIMPLIFIED PITCH ATTITUDE

It has been mentioned previously about different aeroplanes having different spinning characteristics. This is determined by what is called the B/A ratio, and if we refer to Table 1 on page 101, we can see that the ratio compares the moment of inertia about the lateral axis (B) with the moment of inertia about the longitudinal axis (A) as illustrated in Figure 85.

Notes:
1. The moment of inertia about the lateral axis (B) is determined by the positioning of the mass along the length of the fuselage – affecting the pitching moment.
2. The moment of inertia about the longitudinal axis (A) is determined by the positioning of the length of the wings – affecting the rolling moment.

Aerodynamics for the Professional Pilot

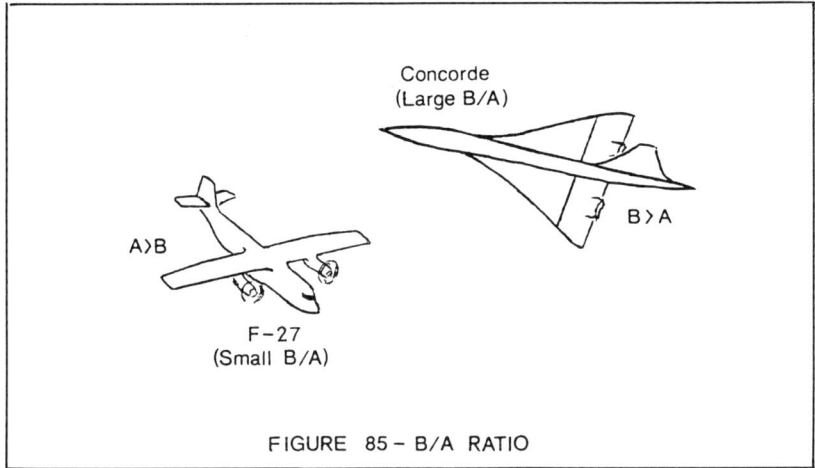

FIGURE 85 – B/A RATIO

Example: In Figure 85 we see that the Concorde has a large B/A ratio (B>A), this meaning that the mass along the length of the fuselage gives a greater turning moment around the lateral axis.

It can be seen also, that the F-27 has a mass allowing greater turning moment, along the length of the wings giving it a small B/A ratio (A>B), around the longitudinal axis.

The B/A can be altered on aircraft dependent upon engine positioning, cargo compartments being full or empty and even aircraft used for parachuting clubs. In this case the mass distribution is altered in flight when the parachutists exit.

This fact is even more pronounced in military tactical transport aircraft such as the Buffalo, Caribou or Hercules C-130. In these instances large masses of freight are carried and can be air-dropped, thereby changing the B/A ratio in flight.

The Recovery

The recovery action has two main aims which are to stop the yaw and to unstall the wings.

As previously mentioned, the distribution of weight between the fuselage and the wings (the B/A ratio) has a large effect on the spin and its recovery action. Although several aircraft have an even distribution of weight between fuselage and wings (which effects inertia and turning moments), the ideal B/A ratio for best anti-spin qualities being 1:3, the B/A ratio can change in flight for all aircraft, due to the ejection of stores and equipment, and/or the fuel consumed.

There is a basic standard method of recovery which covers all aircraft, however some aircraft may have special recovery techniques due to the particular design of the aircraft. These special techniques will not be covered here. The basic standard methods are as follows:

a. Check throttle closed – this is to prevent unnecessary excessive velocity making the recovery more difficult
b. Apply full opposite rudder to direction of spin
c. Move the stick forward until the spin stops, maintaining the ailerons neutral.

Notes:
1. The rudder is normally the primary control but, because the inertial moments are generally large in modern aircraft, aileron deflection is also important.
2. Where the response of the aircraft to rudder is reduced in the spin, the aileron may even be the primary control.
3. With reference to notes 1 and 2, it is in the final analysis that it is the effects of the yawing moment which are the important points and are the effects which make the recovery possible.

d. When the spin stops, centralise the rudder and recover the aircraft from the descent or diving situation using standard techniques (i.e. use of elevators) to level flight.

Inverted Spin

In the inverted spin the motions are more compounded if the aircraft were to follow the same flight path as illustrated in Figure 83 and the changes are described in the following manner and is illustrated in Figure 86.

a. Pitching velocity is in the nose-down sense
b. Rolling velocity is to the right
c. Yawing velocity is to the left

The recovery actions are now more affected as the rolling and yawing motions are in opposite directions. This situation is more particularly so on aircraft with high B/A ratios.

There are basic similarities between the inverted spin and the erect spin, therefore the previously mentioned principles of moment are as equally valid with reference to the inverted spin. Unfortunately there are shielding effects of the inverted spin which are unlike those of the erect spin. This causes the relative airflow over the wing and tail to be altered thereby altering the value of the aerodynamic moments.

Due to the different attitudes of the aircraft in relation to the relative airflow, the control surfaces may have a reduced effect, especially the rudder and the ailerons. The elevators could have the reverse effect and instead of helping recovery (i.e. having anti-spin qualities) it may cause the spin to get worse and therefore more difficult to recover.

To enable the recovery to be effective, deflections of the controls are dependent upon the aircraft's B/A ratio and the direction that the aircraft is yawing, pitching and rolling. The control deflections to be used are:

a. The rudder should be moved to oppose the yaw which is indicated on the turn needle
b. If the B/A ratio is high, then the aileron should be moved in the same direction as the observed roll
c. For conventional aircraft the elevator should be moved up. If the aircraft suffers from shielding effect problems and has a high B/A ratio this control movement may heighten the situation and assist the spinning process.

Shielding Problems
These situations will depend on the position of the wings (i.e. high or low), the tailplane (i.e. high, medium or low) and the fin (i.e. single or multi) with relative positioning to the tailplane itself.

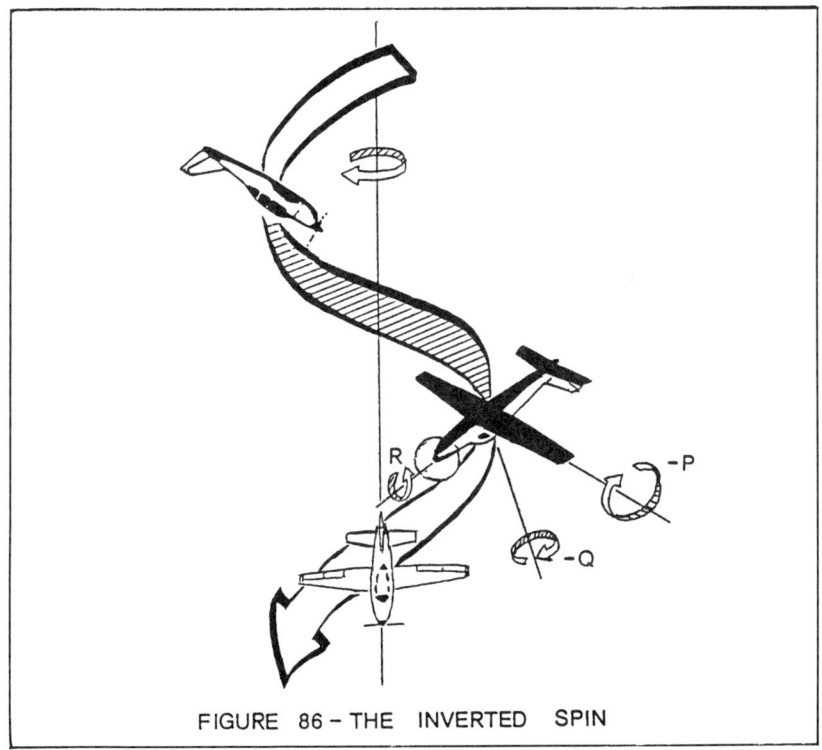

FIGURE 86 – THE INVERTED SPIN